USA and International Mathematical Olympiads 2004

© 2005 by

The Mathematical Association of America (Incorporated)

Library of Congress Catalog Card Number 2005928240

ISBN 0-88385-819-3

Printed in the United States of America

Current Printing (last digit):
10 9 8 7 6 5 4 3 2 1

USA and International Mathematical Olympiads

2004

Edited by

Titu Andreescu   Zuming Feng   Po-Shen Loh

Published and distributed by
The Mathematical Association of America

A Friendly Mathematics Competition: 35 Years of Teamwork in Indiana, edited by Rick Gillman

The Inquisitive Problem Solver, Paul Vaderlind, Richard K. Guy, and Loren C. Larson

International Mathematical Olympiads 1986–1999, Marcin E. Kuczma

Mathematical Olympiads 1998–1999: Problems and Solutions From Around the World, edited by Titu Andreescu and Zuming Feng

Mathematical Olympiads 1999–2000: Problems and Solutions From Around the World, edited by Titu Andreescu and Zuming Feng

Mathematical Olympiads 2000–2001: Problems and Solutions From Around the World, edited by Titu Andreescu, Zuming Feng, and George Lee, Jr.

The William Lowell Putnam Mathematical Competition Problems and Solutions: 1938–1964, A. M. Gleason, R. E. Greenwood, L. M. Kelly

The William Lowell Putnam Mathematical Competition Problems and Solutions: 1965–1984, Gerald L. Alexanderson, Leonard F. Klosinski, and Loren C. Larson

The William Lowell Putnam Mathematical Competition 1985–2000: Problems, Solutions, and Commentary, Kiran S. Kedlaya, Bjorn Poonen, Ravi Vakil

USA and International Mathematical Olympiads 2000, edited by Titu Andreescu and Zuming Feng

USA and International Mathematical Olympiads 2001, edited by Titu
  Andreescu and Zuming Feng

USA and International Mathematical Olympiads 2002, edited by Titu
  Andreescu and Zuming Feng

USA and International Mathematical Olympiads 2003, edited by Titu
  Andreescu and Zuming Feng

USA and International Mathematical Olympiads 2004, edited by Titu
  Andreescu, Zuming Feng, and Po-Shen Loh

MAA Service Center
P. O. Box 91112
Washington, DC 20090-1112
1-800-331-1622   fax: 1-301-206-9789

Contents

Preface

This book is intended to help students preparing to participate in the USA Mathematical Olympiad (USAMO) in the hope of representing the United States at the International Mathematical Olympiad (IMO). The USAMO is the third stage of the selection process leading to participation in the IMO. The preceding examinations are the AMC 10 or the AMC 12 (which replaced the American High School Mathematics Examination) and the American Invitational Mathematics Examination (AIME). Participation in the AIME and the USAMO is by invitation only, based on performance in the preceding exams of the sequence.

The top 12 USAMO students are invited to attend the Mathematical Olympiad Summer Program (MOSP) regardless of their grade in school. Additional MOSP invitations are extended to the most promising non-graduating USAMO students, as potential IMO participants in future years. During the first days of MOSP, IMO-type exams are given to the top 12 USAMO students with the goal of identifying the six members of the USA IMO Team. The Team Selection Test (TST) simulates an actual IMO, consisting of six problems to be solved over two 4 1/2 hour sessions. The 12 equally weighted problems (six on the USAMO and six on the TST) determine the USA Team.

The Mathematical Olympiad booklets have been published since 1976. Copies for each year through 1999 can be ordered from the Mathematical Association of America's (MAA) American Mathematics Competitions (AMC). This publication, as well as Mathematical Olympiads 2000, Mathematical Olympiads 2001, Mathematical Olympiads 2002, and Mathematical Olympiads 2003 are published by the MAA. In addition, various other publications are useful in preparing for the AMC-AIME-USAMO-IMO sequence (see Chapter 6, Further Reading). For more information about the AMC examinations, or to order Mathematical Olympiad booklets from previous years, please write to

Steven Dunbar, MAA Director for K-12 Programs
American Mathematics Competitions
University of Nebraska-Lincoln
1740 Vine Street
Lincoln, NE 68588-0658,

or visit the AMC web site at www.unl.edu/amc.

Acknowledgments

Thanks to Gregory Galperin, Gerald Heuer, Kiran Kedlaya, Ricky Liu, Po-Ru Loh, Cecil Rousseau, and Melaine Wood for contributing problems to this year's USAMO packet. Special thanks to Kiran, Alexander Soifer, Richard Stong, Zoran Sunik, and Melanie for their additional solutions and comments made in their review of the packet. Thanks to Kiran and Richard for their further comments and solutions from grading Problems 2, 5, and 6 on the USAMO. Thanks to Charles Chen, Po-Ru Loh and Tony Zhang who proofread this book. Thanks to Anders Kaseorg, Po-Ru, Tiankai Liu, and Matthew Tang who presented insightful solutions. Also, thanks to Reid Barton, Ricky Liu, and Po-Ru who took the TST in advance to test the quality of the exam.

Abbreviations and Notation

Abbreviations

IMO	International Mathematical Olympiad
USAMO	United States of America Mathematical Olympiad
MOSP	Mathematical Olympiad Summer Program

Notation for Numerical Sets and Fields

| $\mathbb{Z}$ | the set of integers |
| $\mathbb{Z}_n$ | the set of integers modulo n |

Notations for Sets, Logic, and Geometry

$\iff$	if and only if
$\implies$	implies
$|A|$	the number of elements in set A
$A \subset B$	A is a proper subset of B
$A \subseteq B$	A is a subset of B
$A \setminus B$	A without B (the complement of A with respect to B)
$A \cap B$	the intersection of sets A and B
$A \cup B$	the union of sets A and B
$a \in A$	the element a belongs to the set A
AB	the length of segment AB
$\overset{\frown}{AB}$	the arc AB
$\overrightarrow{AB}$	the vector AB

Introduction

Olympiad-style exams consist of several challenging essay-type problems. Correct and complete solutions often require deep analysis and careful argument. Olympiad questions can seem impenetrable to the novice, yet most can be solved by using elementary high school mathematics, cleverly applied.

Here is some advice for students who attempt the problems that follow:

- Take your time! Very few contestants can solve all of the given problems within the time limit. Ignore the time limit if you wish.

- Try the "easier" questions first (problems 1 and 4 on each exam).

- Olympiad problems don't "crack" immediately. Be patient. Try different approaches. Experiment with simple cases. In some cases, working backward from the desired result is helpful.

- If you get stumped, glance at the *Hints* section. Sometimes a problem requires an unusual idea or an exotic technique that might be explained in this section.

- Even if you can solve a problem, read the hints and solutions. They may contain some ideas that did not occur in your solution, and may discuss strategic and tactical approaches that can be used elsewhere.

- The formal solutions are models of elegant presentation that you should emulate, but they often obscure the torturous process of investigation, false starts, inspiration and attention to detail that led to them. When you read the formal solutions, try to reconstruct the thinking that went into them. Ask yourself "What were the key ideas?" "How can I apply these ideas further?"

- Many of the problems are presented together with a collection of remarkable solutions developed by the examination committees, contestants, and experts, during or after the contests. For each problem with multiple solutions, some common crucial results are presented at the beginning of these solutions. You are encouraged to either try to prove those results on your own or to independently complete the solution to the problem based on these given results.

- Go back to the original problem later and see if you can solve it in a different way.

- All terms in boldface are defined in the *Glossary*. Use the glossary and the reading list to further your mathematical education.

- Meaningful problem solving takes practice. Don't get discouraged if you have trouble at first. For additional practice, use prior years' exams or the books on the reading list.

The Problems

1 USAMO

33rd United States of America Mathematical Olympiad

Day I   12:30 PM – 5 PM

April 27, 2004

1. Let $ABCD$ be a quadrilateral circumscribed about a circle, whose interior and exterior angles are at least $60°$. Prove that

$$\frac{1}{3}|AB^3 - AD^3| \le |BC^3 - CD^3| \le 3|AB^3 - AD^3|.$$

When does equality hold?

2. Suppose $a_1, \ldots, a_n$ are integers whose greatest common divisor is 1. Let S be a set of integers with the following properties:

   (a) For $i = 1, \ldots, n, a_i \in S$.

   (b) For $i, j = 1, \ldots, n$ (not necessarily distinct), $a_i - a_j \in S$.

   (c) For any integers $x, y \in S$, if $x + y \in S$, then $x - y \in S$.

   Prove that S must be equal to the set of all integers.

3. For what real values of $k > 0$ is it possible to dissect a $1 \times k$ rectangle into two similar, but incongruent, polygons?

33rd United States of America Mathematical Olympiad

Day II   12:30 PM – 5:00 PM

April 28, 2004

4. Alice and Bob play a game on a 6 by 6 grid. On his or her turn, a player chooses a rational number not yet appearing in the grid and writes it in an empty square of the grid. Alice goes first and then the players alternate. When all squares have numbers written in them, in each row, the square with the greatest number in that row is colored black. Alice wins if she can then draw a line from the top of the grid to the bottom of the grid that stays in black squares, and Bob wins if she can't. (If two squares share a vertex, Alice can draw a line from one to the other that stays in those two squares.) Find, with proof, a winning strategy for one of the players.

5. Let a, b and c be positive real numbers. Prove that

$$(a^5 - a^2 + 3)(b^5 - b^2 + 3)(c^5 - c^2 + 3) \geq (a + b + c)^3.$$

6. A circle ω is inscribed in a quadrilateral $ABCD$. Let I be the center of ω. Suppose that

$$(AI + DI)^2 + (BI + CI)^2 = (AB + CD)^2.$$

Prove that $ABCD$ is an isosceles trapezoid.

2 Team Selection Test

45th IMO Team Selection Test

Lincoln, Nebraska

Day I   **1:00 PM – 5:30 PM**

June 18, 2004

1. Let $a_1, a_2, \ldots, a_n,\ b_1, b_2, \ldots, b_n$ be real numbers such that

$$(a_1^2 + a_2^2 + \cdots + a_n^2 - 1)(b_1^2 + b_2^2 + \cdots + b_n^2 - 1)$$

$$> (a_1 b_1 + a_2 b_2 + \cdots + a_n b_n - 1)^2.$$

   Show that $a_1^2 + a_2^2 + \cdots + a_n^2 > 1$ and $b_1^2 + b_2^2 + \cdots + b_n^2 > 1$.

2. Let n be a positive integer. Consider sequences $a_0, a_1, \ldots, a_n$ such that $a_i \in \{1, 2, \ldots, n\}$ for each i and $a_n = a_0$.

   (a) Call such a sequence *good* if for all $i = 1, 2, \ldots, n$, $a_i - a_{i-1} \not\equiv i \pmod{n}$. Suppose that n is odd. Find the number of good sequences.

   (b) Call such a sequence *great* if for all $i = 1, 2, \ldots, n$, $a_i - a_{i-1} \not\equiv i, 2i \pmod{n}$. Suppose that n is an odd prime. Find the number of great sequences.

3. A 2004×2004 array of points is drawn. Find the largest integer n such that it is possible to draw a convex n-sided polygon whose vertices lie on the points of the array.

45th IMO Team Selection Test

Lincoln, Nebraska

Day II   8:30 AM – 1:00 PM

June 19, 2004

4. Let ABC be a triangle and let D be a point in its interior. Construct a circle ω_1 passing through B and D and a circle ω_2 passing through C and D such that the point of intersection of ω_1 and ω_2 other than D lies on line AD. Denote by E and F the points where ω_1 and ω_2 intersect side BC, respectively, and by X and Y the intersections of lines DF, AB and DE, AC, respectively. Prove that $XY \parallel BC$.

5. Let $A = (0, 0, 0)$ be the origin in the three dimensional coordinate space. The *weight* of a point is the sum of the absolute values of its coordinates. A point is a *primitive lattice point* if all its coordinates are integers with their greatest common divisor equal to 1. A square $ABCD$ is called a *unbalanced primitive integer square* if it has integer side length and the points B and D are primitive lattice points with different weights.

   Show that there are infinitely many unbalanced primitive integer squares $AB_iC_iD_i$ such that the plane containing the squares are not parallel to each other.

6. Let $\mathbb{N}_0^+$ and $\mathbb{Q}$ be the set of nonnegative integers and rational numbers, respectively. Define the function $f : \mathbb{N}_0^+ \to \mathbb{Q}$ by $f(0) = 0$ and

   $$f(3n + k) = -\frac{3f(n)}{2} + k, \quad \text{for } k = 0, 1, 2.$$

   Prove that f is one-to-one, and determine its range.

3   IMO

45th International Mathematical Olympiad

Athens, Greece

Day I   9 AM – 1:30 PM

July 12, 2004

1. Let ABC be an acute triangle with $AB \neq AC$, and let O be the midpoint of segment BC. The circle with diameter BC intersects the sides AB and AC at M and N, respectively. The bisectors of $\angle BAC$ and $\angle MON$ meet at R. Prove that the circumcircles of triangles BMR and CNR have a common point lying on segment BC.

2. Find all polynomials $P(x)$ with real coefficients which satisfy the equality

$$P(a-b) + P(b-c) + P(c-a) = 2P(a+b+c)$$

for all triples (a, b, c) of real numbers such that $ab + bc + ca = 0$.

3. Define a *hook* to be a figure made up of six unit squares as shown in the diagram

or any of the figures obtained by applying rotations and reflections to this figure.

Determine all $m \times n$ rectangles that can be tiled with hooks so that

- the rectangle must be covered without gaps and without overlaps; and

- no part of a hook may cover area outside the rectangle.

45th International Mathematical Olympiad

Athens, Greece

Day II   **9 AM – 1:30 PM**

July 13, 2004

4. Let n be an integer greater than or equal to 3, and let $t_1, t_2, \ldots, t_n$ be positive real numbers such that

$$n^2 + 1 > (t_1 + t_2 + \cdots + t_n)\left(\frac{1}{t_1} + \frac{1}{t_2} + \cdots + \frac{1}{t_n}\right).$$

Show that t_i, t_j, and t_k are side lengths of a triangle for all i, j, and k with $1 \le i < j < k \le n$.

5. In a convex quadrilateral $ABCD$, diagonal BD bisects neither $\angle ABC$ nor $\angle CDA$. Point P lies inside quadrilateral $ABCD$ in such a way that

$$\angle PBC = \angle DBA \quad \text{and} \quad \angle PDC = \angle BDA.$$

Prove that quadrilateral $ABCD$ is cyclic if and only if $AP = CP$.

6. A positive integer is called *alternating* if among any two consecutive digits in its decimal representation, one is even and the other is odd. Find all positive integers n such that n has a multiple which is alternating.

2

Hints

I USAMO

1. Factor $x^3 - y^3$ and apply the **Law of Cosines**.

2. Note the parity argument.

3. It is possible if and only if $k \neq 1$.

4. Bob wins!

5. Find a polynomial $p(x)$ with positive coefficients such that $p(x) \leq x^5 - x^2 + 3$.

6. This is the equality case of an inequality.

2 Team Selection Test

1. Prove the contrapositive or try to find a geometric interpretation.

2. Generating Functions could make your life very easy.

3. Prove that 562 is too big.

4. Find two points concyclic with X, Y, and D.

5. $(2x^2 + 1)^2 = (2x^2 - 1)^2 + (2x)^2 + (2x)^2$.

6. The range is the set of dyadic rational numbers: the rational numbers whose denominators are perfect powers of 2.

3 IMO

1. The common point is the foot of the angle bisector from A.

2. Both x^4 and x^2 work.

3. Find the appropriate coloring(s). You may need more than one.

4. This doesn't require any inequality more high-tech than **AM-GM**.

5. Extend BP and DP to meet the circumcircle of triangle BCD.

6. $20 \nmid n$.

3
Formal Solutions

I USAMO

1. Let $ABCD$ be a quadrilateral circumscribed about a circle, whose interior and exterior angles are at least $60°$. Prove that

$$\frac{1}{3}|AB^3 - AD^3| \le |BC^3 - CD^3| \le 3|AB^3 - AD^3|.$$

When does equality hold?

Solution. By symmetry, we only need to prove the first inequality.

Because quadrilateral $ABCD$ has an incircle, we have $AB + CD = BC + AD$, or $AB - AD = BC - CD$. It suffices to prove that

$$\frac{1}{3}(AB^2 + AB \cdot AD + AD^2) \le BC^2 + BC \cdot CD + CD^2.$$

By the given condition, $60° \le \angle A, \angle C \le 120°$, and so $\frac{1}{2} \ge \cos A, \cos C \ge -\frac{1}{2}$. Applying the **Law of Cosines** to triangle ABD yields

$$BD^2 = AB^2 - 2AB \cdot AD \cos A + AD^2$$
$$\ge AB^2 - AB \cdot AD + AD^2$$
$$\ge \frac{1}{3}(AB^2 + AB \cdot AD + AD^2).$$

The last inequality is equivalent to $(AB - AD)^2 \ge 0$, which is evident. Equality holds if and only if $AB = AD$.

On the other hand, applying the Law of Cosines to triangle BCD yields

$$BD^2 = BC^2 - 2BC \cdot CD \cos C + CD^2 \le BC^2 + BC \cdot CD + CD^2.$$

Combining the above inequalities gives the desired result.

For the equality case, we must have $AB = AD$. This condition is also sufficient, because all the entries in the equalities are 0. Thus, equality holds if and only if $ABCD$ is a **kite** with $AB = AD$ and $BC = CD$.

2. Suppose $a_1, \ldots, a_n$ are integers whose greatest common divisor is 1. Let S be a set of integers with the following properties:

(a) For $i = 1, \ldots, n$, $a_i \in S$.

(b) For $i, j = 1, \ldots, n$ (not necessarily distinct), $a_i - a_j \in S$.

(c) For any integers $x, y \in S$, if $x + y \in S$, then $x - y \in S$.

Prove that S must be equal to the set of all integers.

First Solution. We may as well assume that none of the a_i is equal to 0. We start with the following observations:

(d) $0 = a_1 - a_1 \in S$ by (b).

(e) $-s = 0 - s \in S$ whenever $s \in S$, by (a) and (d).

(f) If $x, y \in S$ and $x - y \in S$, then $x + y \in S$ by (c) and (e).

By (f) plus strong induction on m, we have that $ms \in S$ for any $m \geq 0$ whenever $s \in S$. By (d) and (e), the same holds even if $m \leq 0$, and so we have the following.

(g) For $i = 1, \ldots, n$, S contains all multiples of a_i.

We next verify that

(h) For $i, j \in \{1, \ldots, n\}$ and any integers c_i, c_j, $c_i a_i + c_j a_j \in S$.

We do this by induction on $|c_i| + |c_j|$. If $|c_i| \leq 1$ and $|c_j| \leq 1$, this follows from (b), (d), (f), so we may assume that $\max\{|c_i|, |c_j|\} \geq 2$. Suppose without loss of generality (by switching i with j and/or negating both c_i and c_j) that $c_i \geq 2$; then

$$c_i a_i + c_j a_j = a_i + ((c_i - 1)a_i + c_j a_j)$$

and we have $a_i \in S$, $(c_i-1)a_i + c_j a_j \in S$ by the induction hypothesis, and $(c_i - 2)a_i + c_j a_j \in S$ again by the induction hypothesis. So $c_i a_i + c_j a_j \in S$ by (f), and (h) is verified.

Let e_i be the largest integer such that 2^{e_i} divides a_i; without loss of generality we may assume that $e_1 \geq e_2 \geq \cdots \geq e_n$. Let d_i be the greatest common divisor of $a_1, \ldots, a_i$. We prove by induction on i

that S contains all multiples of d_i for $i = 1, \ldots, n$; the case $i = n$ is the desired result. Our base cases are $i = 1$ and $i = 2$, which follow from (g) and (h), respectively.

Assume that S contains all multiples of d_i, for some $2 \le i < n$. Let T be the set of integers m such that m is divisible by d_i and $m + ra_{i+1} \in S$ for all integers r. Then T contains nonzero positive and negative numbers, namely any multiple of a_i by (h). By (c), if $t \in T$ and s divisible by d_i (so in S) satisfy $t - s \in T$, then $t + s \in T$. By taking $t = s = d_i$, we deduce that $2d_i \in T$; by induction (as in the proof of (g)), we have $2md_i \in T$ for any integer m (positive, negative or zero).

From the way we ordered the a_i, we see that the highest power of 2 dividing d_i is greater than or equal to the highest power of 2 dividing a_{i+1}. In other words, a_{i+1}/d_{i+1} is odd. We can thus find integers f, g with f even such that $fd_i + ga_{i+1} = d_{i+1}$ (choose such a pair without any restriction on f, and replace (f, g) with $(f - a_{i+1}/d_{i+1}, g + d_i/d_{i+1})$ if needed to get an even f.) Then for any integer r, we have $rfd_i \in T$ and so $rd_{i+1} \in S$. This completes the induction and the proof of the desired result.

Second Solution. (By Tony Zhang) We present a different way of completing the proof after observing (d) through (h) of the preceding solution. We proceed to prove the following lemma by induction:

Lemma 1. *Let $m \ge 2$. For $i_1, \ldots, i_m \in \{1, \ldots, n\}$ and $k_{i_1}, \ldots, k_{i_m} \in \mathbb{Z}$:*

$$\left(k_{i_1} a_{i_1} + k_{i_2} a_{i_2}\right) + 2\left(k_{i_3} a_{i_3} + \cdots + k_{i_m} a_{i_m}\right) \in S.$$

Proof: Observation (h) proves the $m = 2$ case, so now assume that the result is true for all m less than or equal to some r. By induction hypothesis, the following terms are in S:

$$\left(k_{i_1} a_{i_1} + k_{i_{r+1}} a_{i_{r+1}}\right) + 2\left(k_{i_3} a_{i_3} + \cdots + k_{i_r} a_{i_r}\right) \in S$$
$$\left(k_{i_2} a_{i_2} + k_{i_{r+1}} a_{i_{r+1}}\right) \in S.$$

Yet their difference is

$$\left(k_{i_1} a_{i_1} - k_{i_2} a_{i_2}\right) + 2\left(k_{i_3} a_{i_3} + \cdots + k_{i_r} a_{i_r}\right),$$

which is in S by the induction hypothesis, so by observation (f), their sum is in S:

$$\left(k_{i_1} a_{i_1} + k_{i_2} a_{i_2}\right) + 2 \left(k_{i_3} a_{i_3} + \cdots + k_{i_r} a_{i_r} + k_{i_{r+1}} a_{i_{r+1}}\right).$$

This completes the induction, and the proof of the Lemma. ∎

We apply Lemma 1 to prove that $1 \in S$; by the comment following observation (f) in the previous solution, this will prove that $S = \mathbb{Z}$. Apply Lemma 1 with $m = n$, $i_1 = 1$, $i_2 = 2$, ..., $i_n = n$. Since we are given that the greatest common divisor of the a_i is 1, there exist integers $k_1, \ldots, k_n$ such that $k_1 a_1 + \cdots + k_n a_n = 1$. Since we can't have all a_i even, without loss of generality, assume that a_1 is odd. Now let i iterate from 2 to n, and at each stage perform the following operations if k_i is odd:

(a) replace k_i with $k_i + a_1$,

(b) replace k_1 with $k_1 - a_i$.

Note that this preserves the sum $k_1 a_1 + \cdots + k_n a_n = 1$, but it makes all the k_i between 2 and n even; therefore, Lemma 1 applies, and we find that $1 \in S$, as desired.

Third Solution. (By Matt Ince) For integers $a_1, \ldots, a_n \in \mathbb{Z}$ with greatest common divisor 1, we say that S *is generated by* $a_1, \ldots, a_n$ if conditions (a), (b), (c) in the problem hold. As in the first solution, we deduce that if S is generated by $a_1, \ldots, a_n$, then

(d) $0 = a_1 - a_1 \in S$ by (b).

(e) $-s = 0 - s \in S$ whenever $s \in S$, by (a) and (d).

Lemma 2. *If S is generated by $a_1, \ldots, a_n$, then S is generated by $a_1, a_2 - a_1, \ldots, a_n - a_1$.*

Proof: Property (c) does not refer to the generators, so we need only check (a) and (b).

(a) We have $a_s - a_1 \in S$ for $s > 1$ by (b).

(b) The difference $a_1 - (a_s - a_1)$ is in S for $s > 1$ by (c), as is its negative by (e). The other differences are of the form $(a_s - a_1) - (a_t - a_1) = a_s - a_t$ for $s, t > 1$, which are in S by (b). ∎

Lemma 3. *If S is generated by $a_1, \ldots, a_n$, then S is generated by $-a_1, a_2, \ldots, a_n$.*

Proof: Again, we need to check the new (a) and (b).

(a) We have $-a_1 \in S$ by (e).

(b) For $s > 1$, we have $a_s - (-a_1) \in S$ by (c) (since $a_s - a_1 \in S$ by (b)), as is its negative by (e). The other differences are in S by (b). ∎

Now suppose S is generated by $a_1, \ldots, a_n$ (and that none of the a_i are zero, without loss of generality); by Lemma 3, we may assume without loss of generality that $a_i > 0$ for each i. Choose integers $b_1, \ldots, b_k > 0$ with greatest common divisor 1 such that S is generated by $b_1, \ldots, b_k$ and $b_1 + \cdots + b_k$ is as small as possible. Note that the b_i must all be distinct (otherwise we could have omitted one), so we may assume without loss of generality that b_1 is smaller than the others.

Suppose $k > 1$, and put $c_1 = b_1$ and $c_s = b_s - b_1$ for $s = 2, \ldots, k$. Then $\gcd(c_1, \ldots, c_k) = \gcd(b_1, \ldots, b_k) = 1$, and S is generated by $c_1, \ldots, c_k$ by Lemma 2. But

$$c_1 + \cdots + c_k = (b_1 + \cdots + b_k) - (k-1)b_1 < b_1 + \cdots + b_k,$$

contradiction. Hence $k = 1$ and $b_1 = 1$.

All that remains is to check that if S is generated by 1, then $S = \mathbb{Z}$. We show that $0, 1, \ldots, k \in S$ for all positive integers k, by induction on k. Note that $-1, 0, 1 \in S$ by (d) and (e), so the base case $k = 1$ is okay. As for the induction step, if $0, 1, \ldots, k \in S$, then $k + 1 = k - (-1) \in S$ by (c). Thus the induction goes through, and all nonnegative integers are in S. By (e), all negative integers are also in S. Hence $S = \mathbb{Z}$, and we are done.

Note. Almost all successful solvers used some variant of the first two solutions. The unique exception was Matt Ince, who received the Clay Mathematics Institute award for coming up with the elegant third solution.

3. For what real values of $k > 0$ is it possible to dissect a $1 \times k$ rectangle into two similar, but incongruent, polygons?

First Solution. We will show that a dissection satisfying the requirements of the problems is possible if and only if $k \neq 1$.

We first show by contradiction that such a dissection is not possible when $k = 1$. Assume that we have such a dissection. The common boundary of the two dissecting polygons must be a single broken line connecting two points on the boundary of the square (otherwise either the square is subdivided into more than two pieces or one of the polygons is inside the other). The two dissecting polygons must have the same number of vertices. They share all the vertices on the common boundary, so they have to use the same number of corners of the square as their own vertices. Therefore, the common boundary must connect two opposite sides of the square (otherwise one of the polygons will contain at least three corners of the square, while the other at most two). However, this means that each of the dissecting polygons must use an entire side of the square as one of its sides, and thus each polygon has a side of length 1. A side of longest length in one of the polygons is either a side on the common boundary or, if all those sides have length less than 1, it is a side of the square. But this is also true of the other polygon, which means that the longest side length in the two polygons is the same. This is impossible since they are similar but not congruent, so we have a contradiction.

We now construct a dissection satisfying the requirements of the problem when $k \neq 1$. Notice that we may assume that $k > 1$, because a $1 \times k$ rectangle is similar to a $1 \times \frac{1}{k}$ rectangle.

We first construct a dissection of an appropriately chosen rectangle (denoted by $ABCD$ below) into two similar incongruent polygons. The construction depends on two parameters (n and r below). By appropriate choice of these parameters we show that the constructed rectangle can be made similar to a $1 \times k$ rectangle, for any $k > 1$. The construction follows.

Let $r > 1$ be a real number. For any positive integer n, consider the following sequence of $2n + 2$ points:

$$A_0 = (0, 0), \quad A_1 = (1, 0), \quad A_2 = (1, r), \quad A_3 = (1 + r^2, r),$$

$$A_4 = (1 + r^2, r + r^3), \quad A_5 = (1 + r^2 + r^4, r + r^3),$$

and so on, until

$$A_{2n+1} = (1 + r^2 + r^4 + \cdots + r^{2n}, r + r^3 + r^5 + \cdots + r^{2n-1}).$$

Define a rectangle $ABCD$ by $A = A_0$, $C = A_{2n+1}$,

$$B = (1 + r^2 + \cdots + r^{2n}, 0), \quad \text{and} \quad D = (0, r + r^3 + \dots + r^{2n-1}).$$

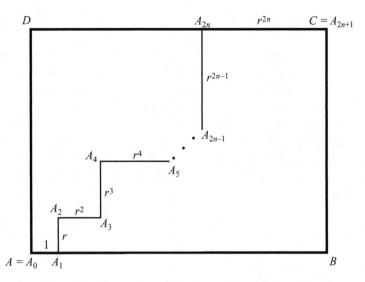

The sides of the $(2n + 2)$-gon $A_1 A_2 \ldots A_{2n+1} B$ have lengths

$$r, \ r^2, \ r^3, \ \ldots, \ r^{2n}, \ r + r^3 + r^5 + \cdots + r^{2n-1}, \ r^2 + r^4 + r^6 + \cdots + r^{2n},$$

and the sides of the $(2n + 2)$-gon $A_0 A_1 A_2 \ldots A_{2n} D$ have lengths

$$1, \ r, \ r^2, \ \ldots, \ r^{2n-1}, \ 1 + r^2 + r^4 + \cdots + r^{2n-2}, \ r + r^3 + r^5 + \cdots + r^{2n-1},$$

respectively. These two polygons dissect the rectangle $ABCD$ and, apart from orientation, it is clear that they are similar but incongruent, with coefficient of similarity $r > 1$. The rectangle $ABCD$ and its dissection are thus constructed.

The rectangle $ABCD$ is similar to a rectangle of size $1 \times f_n(r)$, where

$$f_n(r) = \frac{1 + r^2 + \ldots + r^{2n}}{r + r^3 + \ldots + r^{2n-1}}.$$

It remains to show that $f_n(r)$ can assume any value $k > 1$ for appropriate choices of n and r. Choose n sufficiently large so that $1 + \frac{1}{n} < k$. Since

$$f_n(1) = 1 + \frac{1}{n} < k < k \frac{1 + k^2 + k^4 + \ldots + k^{2n}}{k^2 + k^4 + \ldots + k^{2n}} = f_n(k)$$

and $f_n(r)$ is a continuous function for positive r, there exists an r such that $1 < r < k$ and $f_n(r) = k$, so we are done.

Second Solution. (By Oleg Golberg) We present another proof of the fact that $k = 1$ is impossible. Assume for the sake of contradiction that we have a dissection of a unit square into two polygons that are similar but not congruent. As in the first solution, the dissection must be accomplished via a single path connecting opposite sides of the square. Without loss of generality, suppose that the endpoints of the path are $K \in BC$ and $L \in AD$, where K is a corner of the square if and only if L is the opposite corner. Also, without loss of generality, assume that the right-hand part is strictly smaller than the left-hand part (they are given to be similar but not congruent).

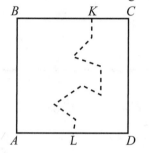

Now, the right-hand part is supposed to be similar to the left-hand part, so let the function F map the right-hand polygon to the left-hand polygon according to the similarity. Observe that the right-hand polygon has the property that if one draws the two perpendiculars to CD at C and D, then these lines completely bound the right-hand polygon. Therefore, after applying F, the same property must hold for $F(CD)$; this must be some side of the left-hand polygon, and its perpendiculars at $F(C)$ and $F(D)$ must bound the entire left-hand polygon. In particular, they must bound A and B. However, there are only a few ways this can be done:

(a) $F(\{C, D\}) = \{A, B\}$. Yet the lengths of CD and AB are equal, so this violates the fact that the similarity is not a congruence.

(b) $F(\{C, D\}) = \{K, L\}$, and $KL \parallel AB$. This has the same problem as the first case.

(c) $F(\{C, D\}) = \{B, K\}$. Since the right-hand polygon is strictly smaller than the left-hand one, this forces $BK > 1 \Rightarrow BC > 1$, contradicting the fact that $ABCD$ is a square.

(d) $F(\{C, D\}) = \{A, L\}$. This has a similar problem to the previous case.

Therefore, all cases yield contradictions, so we have proven $k \neq 1$.

4. Alice and Bob play a game on a 6 by 6 grid. On his or her turn, a player chooses a rational number not yet appearing in the grid and writes it in an empty square of the grid. Alice goes first and then the players alternate. When all squares have numbers written in them, in each row, the square with the greatest number in that row is colored black. Alice wins if she can then draw a line from the top of the grid to the bottom of the grid that stays in black squares, and Bob wins if she can't. (If two squares share a vertex, Alice can draw a line from one to the other that stays in those two squares.) Find, with proof, a winning strategy for one of the players.

First Solution. Bob can win as follows.

After each of his moves, Bob can insure that the maximum number in each row is a square in the set $A \cup B$, where A and B are the sets of squares marked with A and B in the following diagram, respectively.

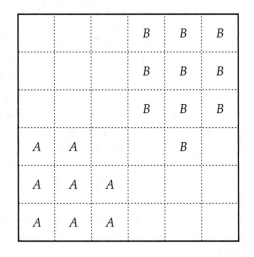

Proof: Bob pairs each square of $A \cup B$ with a square in the same row that is not in $A \cup B$, so that each square of the grid is in exactly one pair. Whenever Alice plays in one square of a pair, Bob will play in the other square of the pair on his next turn. If Alice moves with x in $A \cup B$, Bob writes y with $y < x$ in the paired square. If Alice moves with x not in $A \cup B$, Bob writes z with $z > x$ in the paired square in $A \cup B$. So after Bob's turn, the maximum of each pair is in $A \cup B$, and thus the maximum of each row is in $A \cup B$. ■

So when all the numbers are written, the maximum square in row 1 is in B and the maximum square in row 6 is in A. Since there is no path from B to A that stays in $A \cup B$, Bob wins.

Second Solution. Let P be the property that the following conditions are met:

(a) Row 1, column c is empty if and only if row 2, column $c + 3$ (taken modulo 6) is empty.

(b) If any numbers have been written in the first or second rows, let c_i be the index of the column with the largest number in row i; we require that $|c_1 - c_2| = 3$.

The initial configuration of the grid (empty) clearly satisfies P, and Bob can preserve P by executing the following strategy after each of Alice's moves:

Case 1: Alice moves somewhere in the lower 4 rows. Bob should pick a random square somewhere in the bottom four rows, and write any rational number that has not yet been chosen. This is always possible because the number of squares in the bottom four rows is 24, and if Bob follows this strategy, there will always be an odd number of empty squares (i.e., nonzero) in this section every time Alice's move puts him into this case.

Case 2: Alice moves somewhere in the top two rows. Suppose that she chose the square in row r, column c. If the number she wrote is now the largest number in row r, then Bob should choose a number larger than every number written on the board, and write it in the square at row $3 - r$, column $c + 3$ (taken modulo 6). On the other hand, if Alice's new number is not the largest number in row r, then Bob should choose a number smaller than every number written on the board, and write it in row $3 - r$, column $c + 3$ (taken modulo 6).

It is clear that by the end of Bob's move, P is preserved in both cases. Yet this implies that at the end of the game, $|c_1 - c_2| = 3$, which means that the black squares in the first two rows don't touch anywhere. Therefore, there is no way Alice can connect the top and bottom of the grid with a continuous path, because the blackened region will be disconnected.

5. Let a, b and c be positive real numbers. Prove that

$$(a^5 - a^2 + 3)(b^5 - b^2 + 3)(c^5 - c^2 + 3) \geq (a + b + c)^3.$$

First Solution. For any positive number x, the quantities $x^2 - 1$ and $x^3 - 1$ have the same sign. Thus, we have $0 \leq (x^3 - 1)(x^2 - 1) = x^5 - x^3 - x^2 + 1$, or

$$x^5 - x^2 + 3 \geq x^3 + 2.$$

It follows that

$$(a^5 - a^2 + 3)(b^5 - b^2 + 3)(c^5 - c^2 + 3) \geq (a^3 + 2)(b^3 + 2)(c^3 + 2).$$

It suffices to show that

$$(a^3 + 2)(b^3 + 2)(c^3 + 2) \geq (a + b + c)^3. \qquad (*)$$

We finish with three approaches.

• *First approach* Expanding both sides of inequality $(*)$ and cancelling like terms gives

$$a^3 b^3 c^3 + 3(a^3 + b^3 + c^3) + 2(a^3 b^3 + b^3 c^3 + c^3 a^3) + 8$$
$$\geq 3(a^2 b + b^2 a + b^2 c + c^2 b + c^2 a + a^2 c) + 6abc.$$

By the **AM-GM Inequality**, we have $a^3 + a^3 b^3 + 1 \geq 3a^2 b$. Combining similar results, the desired inequality reduces to

$$a^3 b^3 c^3 + a^3 + b^3 + c^3 + 1 + 1 \geq 6abc,$$

which is evident by the AM-GM Inequality.

• *Second approach* We rewrite the left-hand-side of inequality $(*)$ as

$$(a^3 + 1 + 1)(1 + b^3 + 1)(1 + 1 + c^3).$$

By **Hölder's Inequality**, we have

$$(a^3 + 1 + 1)^{\frac{1}{3}}(1 + b^3 + 1)^{\frac{1}{3}}(1 + 1 + c^3)^{\frac{1}{3}} \geq (a + b + c),$$

from which inequality $(*)$ follows.

• *Third approach* Alternatively, the following double-application of **Cauchy-Schwarz Inequality** also does the trick:

$$\left[(a^3 + 1 + 1)(1 + b^3 + 1)\right]\left[(1 + 1 + c^3)(a + b + c)\right]$$
$$\geq (a^{3/2} + b^{3/2} + 1)^2(a^{1/2} + b^{1/2} + c^2)^2$$
$$\geq (a + b + c)^4.$$

Second Solution. By the AM-GM Inequality,

$$\frac{a^5 + a^5 + 1 + 1 + 1}{5} \geq a^2$$

with equality when $a = 1$. Hence the left hand side of the desired inequality is bounded below by

$$\frac{(3a^5 + 12)(3b^5 + 12)(3c^5 + 12)}{125},$$

so it suffices to prove that

$$(3a^5 + 12)(3b^5 + 12)(3c^5 + 12) \geq 125(a + b + c)^3.$$

After expanding the expressions by brute force, we find that this is equivalent to the following, which we will refer to as (†):

$$27a^5b^5c^5 + 108\sum_{\text{cyc}} a^5b^5 + 432\sum_{\text{cyc}} a^5 + 12^3$$

$$\geq 125\left(\sum_{\text{cyc}} a^3 + 3\sum_{\text{cyc}} a^2b + 3\sum_{\text{cyc}} ab^2 + 6abc\right).$$

Next, we apply AM-GM seven times to derive the following family of inequalities, referred as (‡),

(1) $a^5 + a^5 + a^5 + 1 + 1 \geq 5a^3$, and so

$$25\left[6 + \sum_{\text{cyc}} 3a^5 \geq 5\sum_{\text{cyc}} a^3\right]. \tag{‡}$$

(2) $a^5b^5c^5 + 1 + 1 + 1 + 1 \geq 5abc$, and so

$$27\left[a^5b^5c^5 + 4 \geq 5abc\right]. \tag{‡}$$

(3) $a^5 + a^5b^5 + 1 + 1 + 1 \geq 5a^2b$, and so

$$54 \left[\sum_{\text{cyc}} a^5 + \sum_{\text{cyc}} a^5b^5 + 9 \geq 5 \sum_{\text{cyc}} a^2b \right]. \qquad (\ddagger)$$

Likewise, we have

$$54 \left[\sum_{\text{cyc}} b^5 + \sum_{\text{cyc}} a^5b^5 + 9 \geq 5 \sum_{\text{cyc}} ab^2 \right]. \qquad (\ddagger)$$

(4) $a^5 + b^5 + c^5 + 1 + 1 \geq 5abc$, and so

$$123 \left[a^5 + b^5 + c^5 + 2 \geq 5abc \right]. \qquad (\ddagger)$$

(5) $a^5 + a^5 + b^5 + 1 + 1 \geq 5a^2b$, and so

$$21 \left[\sum_{\text{cyc}} 2a^5 + \sum_{\text{cyc}} b^5 + 6 \geq 5 \sum_{\text{cyc}} a^2b \right]. \qquad (\ddagger)$$

Similarly, we have

$$21 \left[\sum_{\text{cyc}} 2b^5 + \sum_{\text{cyc}} a^5 + 6 \geq 5 \sum_{\text{cyc}} ab^2 \right]. \qquad (\ddagger)$$

Miraculously, the sum of the seven inequalities marked ($\ddagger$) is precisely inequality ($\dagger$), which was what we wanted. The equality condition in each application of the AM-GM Inequality is $a = b = c = 1$, so that is the equality condition for ($\dagger$) as well.

6. A circle ω is inscribed in a quadrilateral $ABCD$. Let I be the center of ω. Suppose that

$$(AI + DI)^2 + (BI + CI)^2 = (AB + CD)^2.$$

Prove that $ABCD$ is an isosceles trapezoid.

Note. We introduce two trigonometric solutions and a synthetic geometry solution. The first solution, by Oleg Golberg, is very technical. The second solution, by Tiankai Liu and Tony Zhang, reveals more geometrical background in their computation. These three students placed in the top three of USAMO 2004. (Indeed, this is by far the most challenging problem in the contest. There are only four complete solutions. The fourth student is Jacob Tsimerman, from

Canada. Evidently, these four students placed top four in the contest.) These three students have won a total of 7 IMO gold medals, with each of Oleg and Tiankai winning 3. Oleg won his first two representing Russia, and the third representing USA. Jacob is one of only four students who scored a perfect paper at the IMO 2004 in Athens, Greece.

The key is to recognize that the given identity is a combination of equality cases of certain inequalities. By equal tangents, we have $AB + CD = AD + BC$ if only if $ABCD$ has an incenter. We will prove that for a convex quadrilateral $ABCD$ with incenter I, then

$$(AI + DI)^2 + (BI + CI)^2 \le (AB + CD)^2 = (AD + BC)^2, \quad (*)$$

and equality holds if and only if $AD \parallel BC$ and $AB = CD$. Without loss of generality, we may assume that the inradius of $ABCD$ is 1.

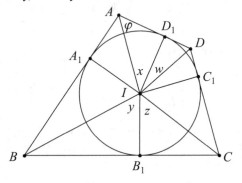

Solution. As shown in the figure above, let A_1, B_1, C_1 and D_1 be the points of tangency. Because circle ω is inscribed in $ABCD$, we can set $\angle D_1 I A = \angle A I A_1 = x$, $\angle A_1 I B = \angle B I B_1 = y$, $\angle B_1 I C = \angle C I C_1 = z$, $\angle C_1 I D = \angle D I D_1 = w$, and $x + y + z + w = 180°$, or $x + w = 180° - (y + z)$, with $0° < x, y, z, w < 90°$. Then $AI = \sec x$, $BI = \sec y$, $CI = \sec z$, $DI = \sec w$, $AD = AD_1 + D_1 D = \tan x + \tan w$, and $BC = BB_1 + B_1 C = \tan y + \tan z$. The inequality $(*)$ becomes

$$(\sec x + \sec w)^2 + (\sec y + \sec z)^2$$
$$\le (\tan x + \tan y + \tan z + \tan w)^2.$$

Expanding both sides of the above inequality and applying the identity

$\sec^2 x = 1 + \tan^2 x$ gives

$$4 + 2(\sec x \sec w + \sec y \sec z)$$
$$\leq 2 \tan x \tan y + 2 \tan x \tan z + 2 \tan x \tan w$$
$$+ 2 \tan y \tan z + 2 \tan y \tan w + 2 \tan z \tan w,$$

or

$$2 + \sec x \sec w + \sec y \sec z$$
$$\leq \tan x \tan w + \tan y \tan z + (\tan x + \tan w)(\tan y + \tan z).$$

Note that by the **Addition-subtraction formulas**,

$$1 - \tan x \tan w = \frac{\cos x \cos w - \sin x \sin w}{\cos x \cos w} = \frac{\cos(x + w)}{\cos x \cos w}.$$

Hence,

$$1 - \tan x \tan w + \sec x \sec w = \frac{1 + \cos(x + w)}{\cos x \cos w}.$$

Similarly,

$$1 - \tan y \tan z + \sec y \sec z = \frac{1 + \cos(y + z)}{\cos y \cos z}.$$

Adding the last two equations gives

$$2 + \sec x \sec w + \sec y \sec z - \tan x \tan w - \tan y \tan z$$
$$= \frac{1 + \cos(x + w)}{\cos x \cos w} + \frac{1 + \cos(y + z)}{\cos y \cos z}.$$

It suffices to show that

$$\frac{1 + \cos(x + w)}{\cos x \cos w} + \frac{1 + \cos(y + z)}{\cos y \cos z} \leq (\tan x + \tan w)(\tan y + \tan z),$$

or

$$s + t \leq (\tan x + \tan w)(\tan y + \tan z),$$

after setting $s = \frac{1 + \cos(x+w)}{\cos x \cos w}$ and $t = \frac{1 + \cos(y+z)}{\cos y \cos z}$. By the Addition-subtraction formulas, we have

$$\tan x + \tan w = \frac{\sin x \cos w + \cos x \sin w}{\cos x \cos w} = \frac{\sin(x + w)}{\cos x \cos w}.$$

Similarly,

$$\tan y + \tan z = \frac{\sin(y + z)}{\cos y \cos z} = \frac{\sin(x + w)}{\cos y \cos z},$$

because $x + w = 180° - (y + z)$. It follows that

$$(\tan x + \tan w)(\tan y + \tan z)$$

$$= \frac{\sin^2(x + w)}{\cos x \cos y \cos z \cos w} = \frac{1 - \cos^2(x + w)}{\cos x \cos y \cos z \cos w}$$

$$= \frac{[1 - \cos(x + w)][1 + \cos(x + w)]}{\cos x \cos y \cos z \cos w}$$

$$= \frac{[1 + \cos(y + z)][1 + \cos(x + w)]}{\cos x \cos y \cos z \cos w} = st.$$

The desired inequality becomes $s + t \le st$, or $(1 - s)(1 - t) = 1 - s - t + st \ge 1$. It suffices to show that $1 - s \le -1$ and $1 - t \le -1$. By symmetry, we only have to show that $1 - s \le -1$; that is,

$$\frac{1 + \cos(x + w)}{\cos x \cos w} \ge 2.$$

Multiplying by $\cos x \cos w$ to both sides of the inequality and applying the Addition and subtraction formulas,

$$1 + \cos x \cos w - \sin x \sin w \ge 2 \cos x \cos w,$$

or, $1 \ge \cos x \cos w + \sin x \sin w = \cos(x - w)$, which is evident. Equality holds if and only $x = w$. Therefore, inequality (∗) is true with equality if and only if $x = w$ and $y = z$, which happens precisely when $AD \parallel BC$ and $AB = CD$, as was to be shown.

Second Solution. We maintain the same notation as in the first solution. Applying the **Law of Cosines** to triangles ADI and BCI,

$$AI^2 + DI^2 = 2\cos(x + w)AI \cdot DI + AD^2$$

and

$$BI^2 + CI^2 = 2\cos(y + z)BI \cdot CI + BC^2.$$

Adding the last two equations and completing squares gives

$$(AI + DI)^2 + (BI + CI)^2 + 2AD \cdot BC$$

$$= 2\cos(x + w)AI \cdot DI + 2\cos(y + z)BI \cdot CI$$

$$+ 2AI \cdot DI + 2BI \cdot CI + (AD + BC)^2$$

Hence, establishing the inequality (∗) is equivalent to establishing the inequality

$$[1 + \cos(x + w)]AI \cdot DI + [1 + \cos(y + z)]BI \cdot CI \leq AD \cdot BC.$$

Since $2[ADI] = AD{\cdot}ID_1 = AI{\cdot}DI \sin(x+w)$, $AI{\cdot}DI = \frac{AD}{\sin(x+w)}$. Similarly, $BI \cdot CI = \frac{BC}{\sin(y+z)}$. Because $x + w = 180° - (y + z)$, we have $\sin(x + w) = \sin(y + z)$ and $\cos(x + w) = -\cos(y + z)$. Plugging all the above information back into the last inequality yields

$$\frac{1 + \cos(x + w)}{\sin(x + w)}AD + \frac{1 - \cos(x + w)}{\sin(x + w)} \cdot BC \leq AD \cdot BC,$$

or

$$\frac{1 + \cos(x + w)}{BC} + \frac{1 - \cos(x + w)}{AD} \leq \sin(x + w). \qquad (∗')$$

Note that, by the Addition-subtraction formulas, the **Product-to-sum formulas**, and the **Double-angle formulas**, we have

$$AD = AD_1 + D_1D = \tan x + \tan w = \frac{\sin x}{\cos x} + \frac{\sin w}{\cos w}$$

$$= \frac{\sin x \cos w + \cos x \sin w}{\cos x \cos w}$$

$$= \frac{\sin(x + w)}{\cos x \cos w} = \frac{2\sin(x + w)}{2\cos x \cos w}$$

$$= \frac{4\sin\frac{x+w}{2}\cos\frac{x+w}{2}}{\cos(x + w) + \cos(x - w)}$$

$$\geq \frac{4\sin\frac{x+w}{2}\cos\frac{x+w}{2}}{\cos(x + w) + 1}$$

$$= \frac{4\sin\frac{x+w}{2}\cos\frac{x+w}{2}}{2\cos^2\frac{x+w}{2}}$$

$$= 2\tan\frac{x + w}{2}.$$

Equality holds if and only if $\cos(x - w) = 1$; that is, $x = w$. (This step can be done easily by applying **Jensen's Inequality**, using the fact that $y = \tan x$ is convex for $0° < x < 90°$.) Consequently, by the Double-angle formulas,

$$\frac{1 - \cos(x + w)}{AD} \leq \frac{2\sin^2 \frac{x+w}{2}}{2\tan \frac{x+w}{2}}$$

$$= \sin\frac{x + w}{2} \cos\frac{x + w}{2}$$

$$= \frac{\sin(x + w)}{2}.$$

In exactly the same way, we can show that

$$\frac{1 + \cos(x + w)}{BC} \leq \frac{\sin(x + w)}{2}.$$

Adding the last equality gives the desired inequality ($*'$). Equality holds if and only if $x = w$ and $y = z$, which happens precisely when $AD \parallel BC$ and $AB = CD$, as was to be shown.

Third Solution. Because circle ω is inscribed in $ABCD$, as shown in the figure, we can set $\angle DAI = \angle IAB = a$, $\angle ABI = \angle IBC = b$, $\angle BCI = \angle ICD = c$, $\angle CDI = \angle IDA = d$, and $a + b + c + d = 180°$. Our proof is based on the following key lemma.

Lemma *If a circle ω, centered at I, is inscribed in a quadrilateral $ABCD$, then*

$$BI^2 + \frac{AI}{DI} \cdot BI \cdot CI = AB \cdot BC. \qquad (\ddagger)$$

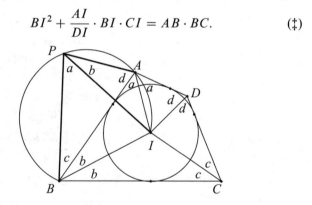

Proof: Construct a point P outside of the quadrilateral such that triangle ABP is similar to triangle DCI. We obtain

$$\angle PAI + \angle PBI = \angle PAB + \angle BAI + \angle PBA + \angle ABI$$

$$= \angle IDC + a + \angle ICD + b$$

$$= a + b + c + d = 180°,$$

implying that the quadrilateral $PAIB$ is cyclic. By **Ptolemy's Theorem**, we have $AI \cdot BP + BI \cdot AP = AB \cdot IP$, or

$$BP \cdot \frac{AI}{IP} + BI \cdot \frac{AP}{IP} = AB. \tag{$\dagger$}$$

Because $PAIB$ is cyclic, it is not difficult to see that, as indicated in the figure, $\angle IPB = \angle IAB = a$, $\angle API = \angle ABI = b$, $\angle AIP = \angle ABP = c$, and $\angle PIB = \angle PAB = d$. Note that triangles AIP and ICB are similar, implying that

$$\frac{AI}{IP} = \frac{IC}{CB} \quad \text{and} \quad \frac{AP}{IP} = \frac{IB}{CB}.$$

Substituting the above equalities into the identity ($\dagger$), we arrive at

$$BP \cdot \frac{CI}{BC} + \frac{BI^2}{BC} = AB,$$

or

$$BP \cdot CI + BI^2 = AB \cdot BC. \tag{$\dagger'$}$$

Note also that triangle BIP and triangle IDA are similar, implying that $\frac{BP}{BI} = \frac{IA}{ID}$, or

$$BP = \frac{AI}{ID} \cdot IB.$$

Substituting the above identity back into ($\dagger'$) gives the desired relation ($*$), establishing the lemma.   ∎

Now we prove our main result. By the lemma and symmetry, we have

$$CI^2 + \frac{DI}{AI} \cdot BI \cdot CI = CD \cdot BC. \tag{$\ddagger'$}$$

Adding the two identities ($\ddagger$) and ($\ddagger'$) gives

$$BI^2 + CI^2 + \left(\frac{AI}{DI} + \frac{DI}{AI} \right) BI \cdot CI = BC(AB + CD).$$

By the **AM-GM Inequality**, we have $\frac{AI}{DI} + \frac{DI}{AI} \geq 2$. Thus

$$BC(AB + CD) \geq IB^2 + IC^2 + 2IB \cdot IC = (BI + CI)^2,$$

where the equality holds if and only if $AI = DI$. Likewise, we have

$$AD(AB + CD) \geq (AI + DI)^2,$$

where the equality holds if and only if $BI = CI$. Adding the last two identities gives the desired inequality ($*$) from the very beginning.

By the given condition in the problem, all the equalities in the above discussion must hold, that is, $AI = DI$ and $BI = CI$. Consequently, we have $a = d$, $b = c$, and so $\angle DAB + \angle ABC = 2a + 2b = 180°$, implying that $AD \parallel BC$. It is not difficult to see that triangle AIB and triangle DIC are congruent, implying that $AB = CD$. Thus, $ABCD$ is an isosceles trapezoid.

2   Team Selection Test

1. Let $a_1, a_2, \ldots, a_n, b_1, b_2, \ldots, b_n$ be real numbers such that

$$(a_1^2 + a_2^2 + \cdots + a_n^2 - 1)(b_1^2 + b_2^2 + \cdots + b_n^2 - 1)$$
$$> (a_1 b_1 + a_2 b_2 + \cdots + a_n b_n - 1)^2.$$

Show that $a_1^2 + a_2^2 + \cdots + a_n^2 > 1$ and $b_1^2 + b_2^2 + \cdots + b_n^2 > 1$.

Note. We present four solutions. The first three are algebraic, proving the contrapositive of the desired statement:

> If $a_1, a_2, \ldots, a_n, b_1, b_2, \ldots, b_n$ are real numbers such that at least one of $a_1^2 + a_2^2 + \cdots + a_n^2$ and $b_1^2 + b_2^2 + \cdots + b_n^2$ is less than or equal to 1, then
>
> $$(a_1^2 + a_2^2 + \cdots + a_n^2 - 1)(b_1^2 + b_2^2 + \cdots + b_n^2 - 1)$$
> $$\leq (a_1 b_1 + a_2 b_2 + \cdots + a_n b_n - 1)^2,$$

or

$$\left(\sum_{i=1}^{n} a_i^2 - 1 \right) \left(\sum_{i=1}^{n} b_i^2 - 1 \right) \leq \left(\sum_{i=1}^{n} a_i b_i - 1 \right)^2 \qquad (*)$$

The fourth solution provides a geometric interpretation.

First Solution. If exactly one of $a_1^2 + a_2^2 + \cdots + a_n^2$ and $b_1^2 + b_2^2 + \cdots + b_n^2$ is less than or equal to 1, then then the left-hand side of the inequality is less than or equal to 0, and inequality $(*)$ is trivial. Now we assume that both are less than or equal to 1. Set $a = 1 - a_1^2 - a_2^2 - \cdots - a_n^2$ and $b = 1 - b_1^2 - b_2^2 - \cdots - b_n^2$. Then both a and b are nonnegative real numbers. Multiplying both sides of the inequality $(*)$ by 4, we reach the equivalent form

$$(2 - 2a_1 b_1 - 2a_2 b_2 - \cdots - 2a_n b_n)^2 \geq 4ab.$$

Note that

$$2 - 2a_1 b_1 - 2a_2 b_2 - \cdots - 2a_n b_n$$
$$= (a_1 - b_1)^2 + (a_2 - b_2)^2 + \cdots + (a_n - b_n)^2 + a + b$$
$$\geq a + b \geq 0.$$

It follows that

$$(2 - 2a_1 b_1 - 2a_2 b_2 - \cdots - 2a_n b_n)^2 \geq (a + b)^2 \geq 4ab,$$

as desired.

Second Solution. Assuming that $a_1^2 + a_2^2 + \cdots + a_n^2$ and $b_1^2 + b_2^2 + \cdots + b_n^2$ are both less than or equal to 1, from the **Cauchy-Schwarz Inequality**, it follows that $a_1 b_1 + a_2 b_2 + \cdots + a_n b_n \leq 1$. Then from the AM-GM Inequality and the above observation above, we deduce that

$$\left(1 - \sum_{i=1}^{n} a_i^2\right)\left(1 - \sum_{i=1}^{n} b_i^2\right) \leq \left(1 - \frac{\sum_{i=1}^{n} a_i^2 + \sum_{i=1}^{n} b_i^2}{2}\right)^2$$

$$\leq \left(1 - \sum_{i=1}^{n} a_i b_i\right)^2,$$

and we are done.

Third Solution. We use the following inequality of Aczel: If $x_1, \ldots, x_m, y_1, \ldots, y_m \in \mathbb{R}$ are such that $x_1^2 > x_2^2 + \cdots + x_m^2$, then

$$(x_1 y_1 - x_2 y_2 - \cdots - x_m y_m)^2 \geq (x_1^2 - x_2^2 - \cdots - x_m^2)(y_1^2 - y_2^2 - \cdots - y_m^2).$$

To prove Aczel's Inequality, we consider the quadratic function

$$f(t) = (x_1 t + y_1)^2 - \sum_{i=2}^{m} (x_i t + y_i)^2$$

and note that $f(-y_1/x_1) \leq 0$. It follows that the discriminant is nonnegative, hence proving the desired inequality. Now we proceed to solve the problem at hand. It is clear that $a_1^2 + \cdots + a_n^2 - 1$ and $b_1^2 + \cdots + b_n^2 - 1$ have the same sign. If both are negative, then Aczel's Inequality yields

$$(1 - a_1 b_1 - \cdots - a_n b_n)^2 \geq (1 - a_1^2 - \cdots - a_n^2)(1 - b_1^2 - \cdots - b_n^2),$$

contradicting the given inequality.

Fourth Solution. Define the vectors $\mathbf{u} = [a_1, a_2, \ldots, a_n]$ and $\mathbf{v} = [b_1, b_2, \ldots, b_n]$, and place $\mathbf{u}$ and $\mathbf{v}$ tail by tail at the origin O to form two sides of a triangle in $\mathbb{R}^n$ by letting A and B be the points with the coordinates of $\mathbf{u}$ and $\mathbf{v}$. The given inequality is equivalent to $(\mathbf{u} \cdot \mathbf{u} - 1)(\mathbf{v} \cdot \mathbf{v} - 1) > (\mathbf{u} \cdot \mathbf{v} - 1)^2$, where "$\cdot$" denotes the **dot product** of two vectors. Expanding both sides and rearranging terms yields

$$(\mathbf{u} \cdot \mathbf{u})(\mathbf{v} \cdot \mathbf{v}) - \mathbf{u} \cdot \mathbf{u} - \mathbf{v} \cdot \mathbf{v} > (\mathbf{u} \cdot \mathbf{v})^2 - 2(\mathbf{u} \cdot \mathbf{v}),$$

or

$$(\mathbf{u} \cdot \mathbf{u})(\mathbf{v} \cdot \mathbf{v}) - (\mathbf{u} \cdot \mathbf{v})^2 > \mathbf{u} \cdot \mathbf{u} + \mathbf{v} \cdot \mathbf{v} - 2(\mathbf{u} \cdot \mathbf{v}).$$

Factoring the right-hand side of the above inequality gives

$$(\mathbf{u} \cdot \mathbf{u})(\mathbf{v} \cdot \mathbf{v}) - (\mathbf{u} \cdot \mathbf{v})^2 > (\mathbf{u} - \mathbf{v}) \cdot (\mathbf{u} - \mathbf{v}).$$

By the vector form of the **Law of Cosines**, we have $(\mathbf{u} \cdot \mathbf{v})^2 = (\mathbf{u} \cdot \mathbf{u})(\mathbf{v} \cdot \mathbf{v}) \cos^2 \angle AOB$. Hence

$$(\mathbf{u} \cdot \mathbf{u})(\mathbf{v} \cdot \mathbf{v})(1 - \cos^2 \angle AOB) > (\mathbf{u} - \mathbf{v}) \cdot (\mathbf{u} - \mathbf{v}),$$

or

$$(\mathbf{u} \cdot \mathbf{u})(\mathbf{v} \cdot \mathbf{v}) \sin^2 \angle AOB > (\mathbf{u} - \mathbf{v}) \cdot (\mathbf{u} - \mathbf{v});$$

that is, $OA^2 \cdot OB^2 \sin^2 \angle AOB > AB^2$. Hence

$$[AOB] = \frac{1}{2} \cdot OA \cdot OB \sin \angle AOB > \frac{1}{2} \cdot AB.$$

Therefore, the length of the altitude from O to AB must be greater than 1. Yet the altitude is the shortest segment connecting O to AB, and so $1 < OA^2 = \mathbf{u} \cdot \mathbf{u} = \sum_{i=1}^{n} a_i^2$ and $1 < OB^2 = \mathbf{v} \cdot \mathbf{v} = \sum_{i=1}^{n} b_i^2$, as desired.

2. Let n be a positive integer. Consider sequences $a_0, a_1, \ldots, a_n$ such that $a_i \in \{1, 2, \ldots, n\}$ for each i and $a_n = a_0$.

   (a) Call such a sequence *good* if for all $i = 1, 2, \ldots, n$, $a_i - a_{i-1} \not\equiv i \pmod{n}$. Suppose that n is odd. Find the number of good sequences.

   (b) Call such a sequence *great* if for all $i = 1, 2, \ldots, n$, $a_i - a_{i-1} \not\equiv i, 2i \pmod{n}$. Suppose that n is an odd prime. Find the number of great sequences.

First Solution. The answer is $(n - 1)^n - (n - 1)$ for part (a) and $(n - 1)((n - 2)^{n-1} - 1)$ for part (b).

(a) Observe that the number of good sequences is clearly the same for any choice of a_0. For fixed a_0, call the condition $a_i - a_{i-1} \not\equiv i \pmod{n}$ *condition (i)*. Now let S_i be the set of sequences $A = \{a_i\}_{i=1}^{n}$, $a_i \in \{1, 2, \ldots, n\}$ such that A satisfies conditions $(1), (2), \ldots, (n - i)$ and fails to satisfy conditions $(n - i + 2), (n - i + 3), \ldots, (n)$. Note that there is no constraint on whether or not A satisfies condition $(n - i + 1)$. Finally, let F be the set of sequences that fail all conditions

$(1), (2), \ldots, (n)$. Then we claim that the number of good sequences starting with a_0 is

$$m = |S_1| - |S_2| + |S_3| - \cdots + |S_n| - |F|.$$

Consider the following table, representing the conditions on sequences in S_i.

	(1)	(2)	(3)	...	$(n-2)$	$(n-1)$	
S_1	(1)	(2)	(3)	...	$(n-2)$	$(n-1)$	
$-\;\;S_2$	(1)	(2)	(3)	...	$(n-2)$		$\overline{(n)}$
$+\;\;S_3$	(1)	(2)	(3)	...		$\overline{(n-1)}$	$\overline{(n)}$
$-\;\;\vdots$	$\vdots$	$\vdots$	$\vdots$	$\ldots$	$\vdots$	$\vdots$	$\vdots$
$+\;\;S_n$		$\overline{(2)}$	$\overline{(3)}$	...	$\overline{(n-2)}$	$\overline{(n-1)}$	$\overline{(n)}$
$-\;\;F$	$\overline{(1)}$	$\overline{(2)}$	$\overline{(3)}$	...	$\overline{(n-2)}$	$\overline{(n-1)}$	$\overline{(n)}$

Conditions that appear as (j) in row i are satisfied by sequences in S_i, and those that appear as $\overline{(j)}$ are failed. Now take any sequence A. If A is good, then A is counted exactly once in $|S_1|$ and is not counted anywhere else. Otherwise, if A is not good, let (i) be the first condition that A fails. If $i > 1$, then either A belongs to exactly the sets S_{n-i+1}, S_{n-i+2} and no others, or A does not belong to any of $S_1, S_2, \ldots, S_n$, and F at all. As one of $|S_{n-i+1}|, |S_{n-i+2}|$ is added and one is subtracted, A has no net contribution to m. Finally, if $i = 1$, then A belongs to S_n and F and no others, so again A does not contribute to m. Therefore, m is indeed the number of good sequences.

We are left with computing the sizes of the S_i's and F. This is easy, as the failure of the final $i - 1$ conditions determines the terms $a_{n-1}, a_{n-2}, \ldots, a_{n-i+1}$, whereas there are just $n - 1$ ways to satisfy each of the conditions $(1), (2), \ldots, (n-i)$, producing $(n-1)^{n-i}$ subsequences $a_0, a_1, \ldots, a_{n-i}$. As there is no constraint on $a_{n-i+1} - a_{n-i}$, we have found all sequences in S_i. Hence, $|S_i| = (n-1)^{n-i}$. As for F, there is clearly exactly one way to fail every single condition, because $1 + 2 + \cdots + (n-1) = \frac{n(n-1)}{2}$ is divisible by n. Therefore, $|S_n| - F = 0$ and

$$m = |S_1| - |S_2| + \cdots - |S_{n-1}|$$
$$= (n-1)^{n-1} - (n-1)^{n-2} + \cdots - (n-1)$$
$$= \frac{(n-1)^n - (n-1)}{n}.$$

As there were originally n choices for a_0, multiplying by n gives the desired answer.

(b)   We can use the same argument as in part (a), but we must recompute the values of $|S_i|$ and $|F|$. Note that for $1 \le i \le n-1$, there are exactly two ways to fail condition (i) (because n is odd) while condition (n) is degenerate and reduces to the single restriction $a_n - a_{n-1} \not\equiv 0 \,(\text{mod } n)$. Hence, we have

$$|S_1| = (n-2)^{n-1} \quad \text{and} \quad |S_i| = 2^{i-2}(n-2)^{n-i},$$

for $2 \le i \le n$. Finding $|F|$ is trickier in this case, however. Considering the differences $b_i = a_i - a_{i-1}$, $i = 1, 2, \ldots, n$, we wish to count the number of choices of $\{b_i\}_{i=1}^{n}$ such that

$$b_1 + b_2 + \cdots + b_n \equiv 0 \quad (\text{mod } n),$$

where $b_i = i$ or $2i \,(\text{mod } n)$ for each i. Note that because of the degenerate condition (n), $b_n \equiv 0$ is forced; hence, we are actually looking at 2^{n-1} sums $b_1 + b_2 + \cdots + b_{n-1}$ rather than 2^n. Now, as $n \mid [1 + 2 + \cdots + (n-1)] = \frac{n(n-1)}{2}$, we can equivalently consider sums of terms $c_i = b_i - i$. That is, we wish to find the number of choices of $c_1, c_2, \ldots, c_{n-1}$ such that

$$c_1 + c_2 + \cdots + c_{n-1} \equiv 0 \quad (\text{mod } n),$$

where $c_i = 0$ or i for each i. We now introduce an extra term c_n that can be either 0 or n. The only effect of this extra term is to double the number of choices of $\{c_i\}_{i=1}^{n}$ compared to the number of choices of $\{c_i\}_{i=1}^{n-1}$. That is, letting

$$F' = \{(c_1, c_2, \ldots, c_n) \mid c_i = 0 \text{ or } i,$$
$$c_1 + c_2 + \cdots + c_n \equiv 0 \,(\text{mod } n)\},$$

we have $|F| = \frac{1}{2}|F'|$.

Observe that we can identify choices $(c_1, c_2, \ldots, c_n)$ with subsets of $T = \{1, 2, \ldots, n\}$; the elements of F' are then simply the subsets of T that have sum divisible by n. The empty set and the set T are two sets that clearly satisfy this property. All remaining subsets $U \subset T$ have from 1 to $n-1$ elements. Write $U + k$ for the set $\{x + k \mid x \in U\}$ (taken mod n); this is the rotation of U by k shifts to the right. Then because n is prime and $0 < |U| < n$, the subsets $U, U+1, U+2, \ldots, U + (n-1)$ form an class of exactly n distinct subsets. Furthermore, the sums of the elements in U and

$U + k$ differ by $k|U| \pmod{n}$, so exactly one element of each class has sum divisible by n. It follows that

$$|F'| = 2 + \frac{2^n - 2}{n}, \quad |F| = \frac{1}{2}|F'| = \frac{2^{n-1} + n - 1}{n}.$$

Therefore,

$$m = |S_1| - |S_2| + |S_3| - \cdots + |S_n| - |F|$$

$$= (n - 2)^{n-1} - (n - 2)^{n-2} + 2(n - 2)^{n-3}$$

$$-2^2(n - 2)^{n-4} + \cdots + 2^{n-2} - \frac{2^{n-1} + n - 1}{n}$$

$$= (n - 2)^{n-1} - \frac{(n - 2)^{n-1} - 2^{n-1}}{n} - \frac{2^{n-1} + n - 1}{n}$$

$$= \frac{(n - 1)(n - 2)^{n-1} - (n - 1)}{n}.$$

As before, multiplying by n gives the desired answer.

Second Solution. We use generating functions in this solution. For part (a), we observe that the number of good sequences is clearly the same for any choice of a_0. Fixing a_0, we represent the choices for the differences $a_i - a_{i-1} \pmod{n}$ with the generating function

$$g_i(x) = 1 + x + x^2 + \cdots + x^{i-1} + x^{i+1} + \cdots + x^{n-1} = p(x) - x^i,$$

where $p(x) = 1 + x + x^2 + \cdots + x^{n-1}$. Then the generating function representing all the possibilities for $a_1 - a_0, a_2 - a_1, \ldots, a_{n-1} - a_{n-2}$ is

$$g(x) = \prod_{i=1}^{n-1} g_i(x) = \prod_{i=1}^{n-1} (p(x) - x^i).$$

The condition that $a_0 - a_{n-1} \not\equiv 0 \pmod{n}$ means that we wish to sum the coefficients of the x^k terms of $g(x)$ for $n \nmid k$.

Now the sum of all the coefficients is simply $g(1) = (n - 1)^{n-1}$. The sum of the coefficients of the x^k terms for $n \mid k$ can be obtained by taking

$$\frac{1}{n} \sum_{j=0}^{n-1} g(\epsilon^j),$$

where $\epsilon = e^{2\pi i/n}$ is a primitive nth root of unity. As we noted above,

$g(1) = (n-1)^{n-1}$. For $n > j > 0$, note that $p(\epsilon^j) = 0$, so

$$g(\epsilon^j) = \prod_{k=1}^{n-1}(-\epsilon)^{kj} = (-1)^{n-1}\epsilon^{(n(n-1)/2)j} = 1,$$

as n is odd. Therefore, the sum of the coefficients of the terms with powers not dividing n is

$$(n-1)^{n-1} - \frac{1}{n}((n-1)^{n-1} + (n-1)) = \frac{1}{n}((n-1)^n - (n-1)).$$

Remembering that we had n choices of a_0 to begin with, we obtain $(n-1)^n - (n-1)$ as the answer to part (a).

For part (b), we change our generating functions to

$$h_i(x) = p(x) - (x^i + x^{2i}), \quad h(x) = \prod_{i=1}^{n-1} h_i(x).$$

Note that for $1 \le i \le n-1$, the terms x^i and x^{2i} both appear because $i \not\equiv 2i \pmod{n}$ as n is odd. For the final condition, however, $n \equiv 2n \equiv 0 \pmod{n}$, so we are still counting those terms—and only those terms—with powers not divisible by n. Hence, using the same method as before, we have $h(1) = (n-2)^{n-1}$,

$$h(\epsilon^j) = \prod_{k=1}^{n-1}(-\epsilon)^{kj}(1 + \epsilon^{kj}) = \prod_{k=1}^{n-1}(1 + \epsilon^{kj}) = \prod_{k=1}^{n-1}(1 + \epsilon^k),$$

for $1 \le j \le n-1$, as n is prime. Now, $1 + \epsilon^i$, $i = 0, 1, \ldots, n-1$, are all of the roots of the polynomial $(x-1)^n - 1$. Therefore, their product is $(-1)^n = -1$ times the constant coefficient of the polynomial, -2. Hence, the product of the roots not including $i = 0$ (in which case $1 + \epsilon^0 = 2$) is 1. Therefore, $h(\epsilon^j) = 1$.

It follows that the number of sequences for a fixed a_0 is

$$(n-2)^{n-1} - \frac{1}{n}((n-2)^{n-1} + (n-1)) = \frac{1}{n}((n-1)((n-2)^{n-1} - 1)),$$

so the answer to part (b) is $(n-1)((n-2)^{n-1} - 1)$.

3. A 2004×2004 array of points is drawn. Find the largest integer n such that it is possible to draw a convex n-sided polygon whose vertices lie on the points of the array.

Note. This is the most challenging problem on this test. No contestant presented a complete solution, and only two students managed to prove that the upper bound was 561; yet both of them failed to construct a 561-gon to finish their solution.

Solution. For a vector $v = (x, y)$, define $\| v \| = |x| + |y|$, the so-called taxicab distance (or taxicab norm). Embed the array of points in the plane such that they correspond to the lattice points in $\{(x, y) : 1 \le x, y \le 2004\}$.

Consider a convex n-gon drawn in our square array, and imagine that we walk along the edges in a counterclockwise direction. Then we can orient each edge and obtain a set of n nonzero vectors $S = \{v_i = (x_i, y_i)\}$, with integer coordinates, whose sum is $(0, 0)$. S has several further properties. First, no two vectors in S are positive multiples of each other by convexity (if $i \ne j$ and the directed edges v_i and v_j are parallel and pointing in the same direction, then our polygon cannot be strictly convex). Second, the sum of the x_i which are positive is at most 2003, and the same is true for the sum of the $y_i > 0$, as well as the sums of the $|x_i|$ and $|y_i|$ for $x_i < 0$ and $y_i < 0$. This is true because all the vectors with, say, x_i positive will correspond to adjacent edges (by convexity), and if one traces these edges in order on the polygon, one must start from a point within $\{(x, y) : 1 \le x, y \le 2004\}$ and finish at a point in that same region; therefore the total displacement in the x-dimension is bounded by 2003. In particular, this implies that $\sum \| v_i \| \le 8012$. Third, given S that satisfies the above properties, we can construct a convex polygon that fits within the bounds: since the polygon should be convex, we must place the vectors end to end ordered by the angle (measured counterclockwise) that they make with the positive x-axis, and the resulting polygon will fit within the array because of the given inequalities.

We now show that it is impossible to draw a 562-gon in the array. We will prove that for every set S of 562 vectors such that no two are positive multiples of each other, $\sum \| v_i \| > 8012$. We can even ignore the condition that $\sum v_i = (0, 0)$. Let us now try to minimize $\sum \| v_i \|$. Since no two vectors in S are positive multiples of each other, we may assume that for each i, $\gcd(|x_i|, |y_i|) = 1$, or else we might as well scale that vector down by that gcd (it will only reduce $\sum \| v_i \|$). Now:

Lemma   *For any positive integer k, the maximum number of distinct $v_i = (x_i, y_i)$ that satisfy $\| v_i \| = k$ and $\gcd(|x_i|, |y_i|) = 1$ is $4\phi(k)$, where $\phi(k)$ denote Euler's totient function.*

Proof: For $k = 1$, the lemma is trivial. For $k \geq 2$, the gcd condition forces that neither of x_i or y_i can ever be 0. So, it suffices to show that the number of v_i with $x_i > 0$ and $y_i > 0$ is $\phi(k)$, because there are exactly 4 ways to choose pairs of signs, for the coordinates of each v_i. Because $\|v_i\| = k$, this means that $|x_i| + |y_i| = k$. Hence $gcd(|x_i|, |y_i|) = 1$ if and only if $gcd(|x_i|, k) = 1$. By definition of $\phi(k)$, there are precisely $\phi(k)$ ways to choose positive x_i to satisfy this condition; each way yields exactly one positive choice for y_i, so we are done with the proof of this lemma.   ∎

Now for $n = 562$, since $\sum_{k=1}^{21} 4\phi(k) = 560$, we minimize $\sum \|v_i\|$ when we use all 560 vectors with $\|v_i\| \leq 21$, and 2 more vectors with $\|v_i\| = 22$. The value of this sum is

$$\sum \|v_i\| \geq \sum_{k=1}^{21} 4k\phi(k) + 2 \cdot 22 = 8032 > 8012,$$

as was claimed. Thus $n \leq 561$.

Indeed, we now construct a 561-gon in the array by constructing a suitable set S. Start by setting S to be the set of all 560 integer vectors (x_i, y_i) with norm at most 21 and $gcd(|x_i|, |y_i|) = 1$. By symmetry, the vector sum of these 560 vectors is $(0, 0)$. Also by symmetry, the sum of all of the positive x_i is exactly $1/4$ of the sum of all of the taxicab norms, so it is $\sum_{k=1}^{21} k\phi(k) = 1997$. This is the span of the 560-gon in the x-dimension. Similarly, that is also the span of the 560-gon in the y-dimension.

Next, remove from S the vectors $(20, 1)$ and $(-11, 10)$, and insert the vectors $(5, 17)$, $(-17, -5)$, and $(21, -1)$. Now we have a 561-gon. Since the vector sum of $(20, 1)$ and $(-11, 10)$ is equal to the vector sum of the 3 new vectors, this preserves the fact that $\sum v_i = (0, 0)$, so we can still have a closed, convex polygon. Let us see what has happened to the x-span and y-span: we removed 20 from the sum of positive x_i, but added back 5 and 21. This is a net gain of 6, so the sum of positive x_i will increase from 1997 to 2003—this exactly matches the upper limit imposed by the second observation at the beginning of the proof. Similarly, the sum of positive y_i will lose summands of 1 and 10 from the deletion, but it will gain 17 from the insertion. This is also a net gain of 6, so we also have 2003 as the final sum of positive y_i. Since the vector sum of all v_i is $(0, 0)$, similar results with 2003 hold for the negative x_i and y_i. By the third observation at the beginning of the proof, we are done.

4. Let ABC be a triangle and let D be a point in its interior. Construct a circle ω_1 passing through B and D and a circle ω_2 passing through C and D such that the point of intersection of ω_1 and ω_2 other than D lies on line AD. Denote by E and F the points where ω_1 and ω_2 intersect side BC, respectively, and by X and Y the intersections of lines DF, AB and DE, AC, respectively. Prove that $XY \parallel BC$.

First Solution. Let circles ω_1 and ω_2 meet again at R (other than D), and let ω_1 and ω_2 intersect again with segments AB and AC, respectively, at P and Q (other than B and C). By the **Power of a Point Theorem**, we have $AP \cdot AB = AR \cdot AD = AQ \cdot AC$; that is, points A, R, and D lie on the **radical axis** of the circles ω_1 and ω_2. Therefore points $B, C, Q,$ and P lie on a circle. Consequently, $\angle AQP = \angle ABC$. To prove that $XY \parallel BC$, it suffices to show that $\angle AXY = \angle ABC$, or $\angle AQP = \angle ABC = \angle AXY$; that is, to prove that $P, Q, Y,$ and X lie on a circle.

Because $BDRP$ is cyclic, $\angle PDY = \angle PBE = \angle ABC$. Hence $\angle AQP = \angle PDY$, implying that P, Q, Y and D lie on a circle. Similarly, we can show that $P, Q, D,$ and X lie on a circle. We conclude that $P, Q, Y, D,$ and X all lie on the circumcircle of triangle PQD, completing the proof.

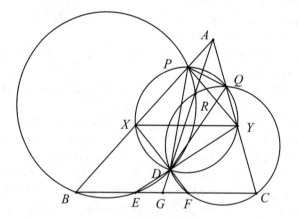

Second Solution. It suffices to show that

$$\frac{AX}{XB} = \frac{AY}{YC}. \qquad (*)$$

Let line AD and segment BC meet at G. Applying **Menelaus's Theorem** to triangle ABG and line XF, triangle ACG and line YE gives

$$\frac{AX \cdot BF \cdot GD}{XB \cdot FG \cdot DA} = 1 \text{ and } \frac{AY \cdot CE \cdot GD}{YC \cdot EG \cdot DA} = 1.$$

Hence

$$\frac{AX \cdot BF}{XB \cdot FG} = \frac{AY \cdot CE}{YC \cdot EG}.$$

To establish the equation (∗), it suffices to show that

$$\frac{BF}{FG} = \frac{CE}{EG},$$

or $EG \cdot BF = CE \cdot FG$. By the Power of a Point Theorem, we have $EG \cdot BG = GD \cdot GR = GF \cdot GC$, implying that

$$EG \cdot BF = EG(BG + GF) = EG \cdot BG + EG \cdot GF$$

$$= GF \cdot GC + EG \cdot GF = GF(GC + GE)$$

$$= CE \cdot FG,$$

as desired.

Note. The result still holds when points E and F lie on line BC instead of segment BC. We added the restriction of segment only to make the arguments on configurations easier for the students taking the test.

5. Let $A = (0, 0, 0)$ be the origin in the three dimensional coordinate space. The *weight* of a point is the sum of the absolute values of its coordinates. A point is a *primitive lattice point* if all its coordinates are integers with their greatest common divisor equal to 1. A square $ABCD$ is called a *unbalanced primitive integer square* if it has integer side length and the points B and D are primitive lattice points with different weights.

   Show that there are infinitely many unbalanced primitive integer squares $AB_iC_iD_i$ such that the planes containing the squares are not parallel to each other.

Note. This problem turned out to be the second most challenging problem on the test after problem 3. There were three complete solutions. Two of those solutions used a simple algebraic identity

$$(2x^2 + 1)^2 = 4x^4 + 4x^2 + 1 = (2x^2 - 1)^2 + 4x^2 + 4x^2.$$

This identity leads to the second solution shown below. We provide some motivation for this idea in the first solution.

First Solution. Let (a, b, c) be a **Pythagorean Triple**, that is, a, b, c are positive integers with $a^2 + b^2 = c^2$. The key facts are:

(i) for arbitrary positive integers m and n such that $m > n$, $(m^2 - n^2, 2mn, m^2 + n^2)$ is a Pythagorean triple; and

(ii) the vector $[c - a, c - b, a + b - c]$ has integer length $2c - a - b$.

Fact (i) is rather elementary. To establish (ii), we compute

$$(c - a)^2 + (c - b)^2 + (a + b - c)^2 - (2c - a - b)^2$$
$$= 2c^2 + a^2 + b^2 - 2c(a + b) + (a + b)^2 - 2c(a + b) + c^2$$
$$\quad -4c^2 + 4c(a + b) - (a + b)^2$$
$$= a^2 + b^2 - c^2 = 0.$$

Putting these two facts together demonstrates that the vector $[2n^2, (m - n)^2, 2n(m - n)]$ has integer length. For $i = 1, 2, \ldots$, let $o_i = 2i + 1$. We set $n = 1$ and $m - n = o_i$, and define vectors

$$\overrightarrow{AB_i} = \mathbf{u}_i = [2, 2o_i, o_i^2], \quad \overrightarrow{AD_i} = \mathbf{v}_i = [2o_i, o_i^2 - 2, -2o_i],$$

and

$$\mathbf{w}_i = [-o_i^2, 2o_i, -2].$$

It is not difficult to check that $2^2 + (2o_i)^2 + (o_i^2)^2 = (o_i^2 + 2)^2$, and that

$$(2o_i)^2 + (o_i^2 - 2)^2 + (-2o_i)^2$$
$$= 8o_i^2 + o_i^4 - 4o_i^2 + 4$$
$$= o_i^4 + 4o_i^2 + 4 = (o_i^2 + 2)^2,$$

implying that the vectors $\mathbf{u}_i, \mathbf{v}_i, \mathbf{w}_i$ have the same integer length. Because $\gcd(2, o_i) = \gcd(2o_i, o_i^2 - 2) = 1$, points $B_i = (2, 2o_i, o_i^2)$ and $D_i = (2o_i, o_i^2 - 2, -2o_i)$ are primitive lattice points. It is also routine to check that the weights of B_i and D_i are distinct, because $2 + 2o_i + o_i^2 \neq 4o_i + o_i^2 - 2$, that is, $2 \neq o_i$. Finally, we claim that $AB_i \perp AD_i$, because the **dot product**

$$\mathbf{u}_i \cdot \mathbf{v}_i = 4o_i + 2o_i(o_i^2 - 2) - 2o_i^3$$
$$= 4o_i + 2o_i^3 - 4o_i - 2o_i^3 = 0.$$

It follows that there are infinitely many unbalanced primary integer squares $AB_iC_iD_i$. (We can set $C_i = B_i + \mathbf{v}_i$.)

It suffices to show that the planes containing $AB_iC_iD_i$ are not parallel to each other. We show that the normal vectors of the planes containing $AB_iC_iD_i$ are not parallel. We first claim that $\mathbf{w}_i$ is the normal vector of the plane containing $AB_iC_iD_i$, because

$$\mathbf{w}_i \cdot \mathbf{u}_i = -2o_i^2 + 4o_i^2 - 2o_i^2 = 0,$$

$$\mathbf{w}_i \cdot \mathbf{v}_i = -2o_i^3 + 2o_i(o_i^2 - 2) + 4o_i$$

$$= 2o_i(-o_i^2 + o_i^2 - 2 + 2) = 0.$$

Finally, it is not difficult to see that the $\mathbf{w}_i$ are not parallel, because no two are related by a scalar multiplication.

Second Solution. We maintain the same notation as in the first solution. For $i = 1, 2, \ldots$, let $e_i = 2i$. We set $n = 1$ and $m - n = e_i$. Then $[2n^2, (m-n)^2, 2n(m-n)] = [2, 4i^2, 4i] = 2[1, 2i^2, 2i]$. We define vectors

$$\overrightarrow{AB_i} = \mathbf{u}_i' = [1, 2i, 2i^2], \quad \overrightarrow{AD_i} = \mathbf{v}_i' = [-2i, 1 - 2i^2, 2i],$$

and

$$\mathbf{w}_i' = [-2i^2, 2i, -1].$$

For each i, the vectors $(\mathbf{u}_i, \mathbf{v}_i, \mathbf{w}_i)$ and $(\mathbf{u}_i', \mathbf{v}_i', \mathbf{w}_i')$ are dual to each other. It is straightforward to check that pairs of vectors $(\mathbf{u}_i', \mathbf{v}_i')$ provides another family of infinitely many unbalanced primitive integer squares $AB_iC_iD_i$ such that the planes containing the squares are not parallel to each other.

Third Solution. Let (a, b, c) be a primitive Pythagorean triple, that is, a, b, c are positive integers with $a < b < c$, $a^2 + b^2 = c^2$ and $\gcd(a, b, c) = 1$. It is not difficult to show that $\gcd(a, b) = \gcd(c, b) = \gcd(c, a) = 1$. In order to find a vector with integer length, we consider $c^4 = c^2(a^2 + b^2) = c^2a^2 + c^2b^2 = (a^2 + b^2)a^2 + c^2b^2 = c^2a^2 + (a^2 + b^2)b^2$. Set

$$\overrightarrow{AB} = \mathbf{u} = [-ab, b^2, ac] \quad \text{and} \quad \overrightarrow{AD} = \mathbf{v} = [a^2, -ab, bc].$$

Also set $C = (a^2 - ab, b^2 - ab, c(a + b))$.

By construction, $AB^2 = AD^2 = (a^2 + b^2)c^2 = c^4$ and

$$\overrightarrow{AB} \cdot \overrightarrow{AD} = \mathbf{u} \cdot \mathbf{v} = ab(-a^2 - b^2 + c^2) = 0.$$

We also note that the difference of the weights of B and D is equal to

$$(b^2 + ab + ac) - (a^2 + ab + bc)$$
$$= b^2 - a^2 + ac - bc$$
$$= (b - a)(a + b) + c(a - b)$$
$$= (b - a)(a + b - c),$$

which is nonzero because $\gcd(a, b) = 1$ and $a, b,$ and c are sides of a triangle. Because $\gcd(a, b) = \gcd(c, b) = \gcd(c, a) = 1$, $\gcd(b^2, ac) = \gcd(a^2, bc) = 1$. Therefore, $ABCD$ is a unbalanced primitive integer square (associated to this primitive Pythagorean triple).

We note that vector $\mathbf{w} = [b, a, 0]$ is perpendicular to the plane containing the square $ABCD$. Indeed,

$$\mathbf{u} \cdot \mathbf{w} = -ab^2 + ab^2 = 0 = a^2b - a^2b = \mathbf{v} \cdot \mathbf{w}.$$

Finally, it is not difficult to see that for distinct primitive Pythagorean triples (a, b, c) and (a', b', c'), the vectors $\mathbf{w} = [b, a, 0]$ and $\mathbf{w}' = [b', a', 0]$ are not parallel to each other. Hence, the unbalanced primitive integer squares associated with distinct primitive Pythagorean triples lie in different planes. It is well known that there are infinitely many primitive Pythagorean triples, so our proof is complete.

6. Let $\mathbb{N}_0^+$ and $\mathbb{Q}$ be the set of nonnegative integers and rational numbers, respectively. Define the function $f : \mathbb{N}_0^+ \to \mathbb{Q}$ by $f(0) = 0$ and

$$f(3n + k) = -\frac{3f(n)}{2} + k, \quad \text{for } k = 0, 1, 2.$$

Prove that f is one-to-one, and determine its range.

Solution. We prove that the range of f is the set T of rational numbers of the form $m/2^n$ for $m \in \mathbb{Z}$ and $n \in \mathbb{N}_0^+$ (also known as the *dyadic rational numbers*). For $x, y \in T$, we write $x \equiv y \pmod 3$ to mean that the numerator of $x - y$, when written in lowest terms, is divisible by 3. Define the function $g : T \to \{0, 1, 2\}$ by declaring that for $x \in T$, $g(x)$ is the unique element of $\{0, 1, 2\}$ such that $g(x) \equiv x \pmod 3$.

We first check that f is one-to-one. Suppose that $f(a) = f(b)$ for some $a, b \in \mathbb{N}_0^+$ with $a < b$; choose such a pair with b minimal.

Now note that $f(a) \equiv a$ (mod 3) and $f(b) \equiv b$ (mod 3) from the definition of f. Hence $a \equiv b$ (mod 3). Choose the $i \in \{0, 1, 2\}$ that is congruent to a and b modulo 3, and put $a' = (a - i)/3$ and $b' = (b - i)/3$. Then $f(a') = f(b')$, but $a' < b'$ and $b' < b$ since $b > 0$. This contradicts the choice of a and b. Hence no such pairs exist.

We next verify that T, which clearly contains the range of f, is in fact equal to it. Define the map $h : T \rightarrow T$ by setting $h(x) = \frac{2}{3}(g(x) - x)$. Then x is in the image of f whenever $h(x)$ is: if $h(x) = f(a)$, then $x = f(3a + g(x))$. Hence it suffices to show that if one starts from x and applies h repeatedly, one eventually ends up with an element of the range of f.

First note that if x is not an integer, then $g(x) - x$ has the same denominator as x, and $h(x)$ has denominator half of that. Hence some x_i is an integer.

Next, by the **Triangle Inequality**,

$$|h(x)| \le \frac{2}{3}(2 + |x|) < |x|$$

whenever $|x| > 4$. Thus, our process eventually hits an integer of absolute value bounded by 4. Simple computation yields the following table:

x	-4	-3	-2	-1	0	1	2	3	4
$h(x)$	4	2	2	2	0	0	0	-2	-2

From the table, it is clear that starting from any point in $\{-4, -3, \ldots, 4\}$, within 4 applications of h, our process hits zero. Since $f(0) = 0$, this value is in the range of f, so we are done.

Note. It is of interest to observe that the function f can be interpreted as follows: to compute $f(n)$, first write n in base 3, and then read off the digit string as if it were in base $(-3/2)$. Therefore, the point of this question is to prove that all dyadic rational numbers can be uniquely expressed in base $(-3/2)$.

3 IMO

1. Let ABC be an acute triangle with $AB \neq AC$, and let O be the midpoint of segment BC. The circle with diameter BC intersects the sides AB and AC at M and N, respectively. The bisectors of $\angle BAC$ and $\angle MON$ meet at R. Prove that the circumcircles of triangles BMR and CNR have a common point lying on segment BC.

Note. We present three solutions. Extend segment AR through R to meet side BC at D (that is, line AD bisects $\angle BAC$). We will prove that the circumcircles of triangles BMR and CNR meet at D.

First Solution. We start with a well-known geometric fact.

Lemma. *Let AMN be a triangle with $AM \neq AN$. Then the intersection of the perpendicular bisector of segment MN and the bisector of $\angle MAN$ meet at a point R on the circumcircle of triangle AMN.*

Proof: Because $AM \neq AN$, point R is uniquely defined. Let ω denote the circumcircle of triangle AMN. It is not difficult to see that the midpoint of the arc MN (not including A) is the point R sought. ∎

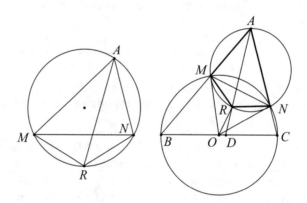

Now we prove our main problem. Since quadrilateral $BCNM$ is cyclic,

$$\angle AMN = \angle ACB \quad \text{and} \quad \angle ANM = \angle ABC. \tag{$*$}$$

Hence triangles ABC and ANM are similar. Since $AB \neq AC$, $AN \neq AM$. Because OMN is an isosceles triangle with $OM = ON$, the bisector of $\angle MON$ is also the perpendicular bisector of MN. Applying the lemma to triangle AMN, we conclude that R

lies on the circumcircle of triangle AMN; that is, $AMRN$ is cyclic. It follows that $\angle ARM = \angle ANM$. By the second equality in $(*)$, we obtain $\angle ARM = \angle ABC$. Extend segment AR through to meet side BC at D. Then $\angle ARM = \angle ABD$; that is, $BDRM$ is cyclic. Likewise, we can show that $CDRN$ is cyclic. Therefore, the circumcircles of triangles BMR and CNR meet at D, as desired.

Second Solution.

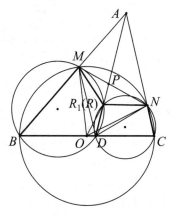

Note that $\angle AMC = \angle ANB = 90°$ and that $\angle AMD > \angle AMC$ and $\angle AND > \angle ANB$. Hence $\angle AMD + \angle AND > 180°$. In quadrilateral $AMDN$,

$$\angle MDA + \angle NDA = \angle MDN$$
$$= 360° - \angle BAC - \angle AMD - \angle AND$$
$$< 180° - \angle BAC = \angle B + \angle C.$$

Let segments AD and MN meet at P. Then $\angle MPA + \angle NPA = \angle MPN = 180° > \angle B + \angle C$. Let R_1 be a point moving along segment PD from D to P; then there is a position for R_1 such that at least one of $\angle MR_1 A = \angle B$ and $\angle NR_1 A = \angle C$ is true. Without loss of generality, we assume that

$$\angle MR_1 A = \angle B;$$

that is, quadrilateral $BDR_1 M$ is cyclic. Note also that $BMNC$ us cyclic. By the **Power of a Point Theorem**, we have

$$AR_1 \cdot AD = AM \cdot AB = AN \cdot AC,$$

implying that quadrilateral CDR_1N is cyclic. It follows that $\angle AR_1N = \angle C$.

Therefore, we have $\angle MR_1N = \angle MR_1A + \angle AR_1N = \angle B + \angle C$. Consequently, $\angle MAN + \angle MR_1N = \angle BAC + \angle B + \angle C = 180°$; that is, quadrilateral AMR_1N is cyclic. By the **Extended Law of Sines**, it follows that

$$\frac{MR_1}{\sin \angle MAR_1} = \frac{NR_1}{\sin \angle NAR_1}.$$

Note that $\angle MAR_1 = \angle BAD = \angle CAD = \angle NAR_1$. We conclude that $MR_1 = NR_1$; that is, R_1 lies on the perpendicular bisector of segment MN. Note that O also lies on the perpendicular bisector of segment MN. Hence line OR_1 is the perpendicular bisector of segment MN. Note also that R_1 lies inside triangle ABC. We conclude that ray MR_1 bisects $\angle MON$; that is, $R_1 = R$.

Because all of the above constructions are unique, we conclude that the circumcircles of triangles BMR and CNR meet at D.

Third Solution. Without loss of generality, we assume that $AB > AC$. We have the following configuration.

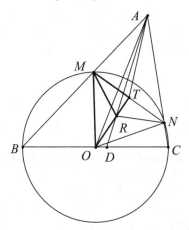

Denote by T the intersection of OR and MN. Because $OM = ON$, T is the midpoint of segment MN. As shown in the second proof, triangles AMN and ACB are similar. Because AT and AO are the corresponding medians of similar triangles AMN and ACB, $\angle BAO = \angle CAT$. Note that $\angle BAD = \angle CAD$. We conclude that $\angle OAR = \angle RAT$; that is, line AR also bisects $\angle OAT$. By the **Angle**

Bisector Theorem, we have

$$\frac{RT}{RO} = \frac{AT}{AO}.$$

Furthermore, using again the fact AT and AO are the corresponding medians in similar triangles AMN and ACB, we obtain

$$\frac{AT}{AO} = \frac{MN}{BC} = \frac{MT}{BO} = \frac{MT}{MO}.$$

It follows that $\frac{RT}{RO} = \frac{MT}{MO}$, implying that line MR bisects $\angle TMO = \angle NMO$. Because $BCNM$ is cyclic, $\angle AMN = \angle ACB$. Because $OB = OM$, $\angle BMO = \angle ABC$. Hence $\angle NMO = 180° - \angle AMN - \angle BMO = 180° - \angle ABC - \angle ACB = \angle BAC$, and so $\angle OMR = \frac{\angle OMN}{2} = \frac{\angle BAC}{2}$. Therefore,

$$\angle BMR = \angle BMO + \angle OMR = \angle ABC + \frac{\angle BAC}{2} = \angle ADC,$$

implying that $BDRM$ is cyclic. Likewise, we can show that $CDRN$ is cyclic.

Note. Compare the second solution to the first solution of TST problem 4. The USA IMO delegation pointed out the similarity of these two approaches during the IMO 2004 jury meetings. Most team leaders believed that the problems contained sufficiently distinct geometric ingredients and chose the problem to be on the exam. By the way, this problem is a good twist of a simple version of **Miquel's Theorem**:

> Let ABC be a triangle. Points $X, Y,$ and Z lie on sides $BC, CA,$ and AB, respectively. The circumcircles of triangles $AYZ, BZX,$ and CXY meet at a common point.

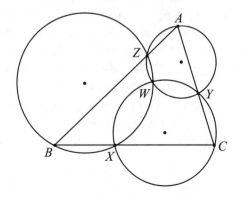

The third solution may seem only remotely connected to the first two solutions. The following might help the reader to tie the ideas closer together.

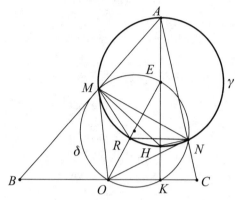

Let K be the foot of the perpendicular from A to BC. Then altitudes AK, BN, and CM meet at the orthocenter H. Let E be the midpoint of segment AH. Then the circumcircle of triangle KMN is δ, the 9-point circle of triangle ABC; that is, δ passes through the points K, M, N, and all midpoints of segments AB, BC, CA, AH, BH, and CH. (There are many ways to prove the existence of the 9-point circle of a triangle. One simple way is to recognize that the center of the 9-point circle is halfway between the circumcenter and the orthocenter of the triangle. We leave the details to the reader.) Then, OE is a diameter of δ and the symmetry axis of the **kite** $OMEN$. Because $\angle ANH = \angle AMH = 90°$, quadrilateral $AMHN$ is inscribed in a circle γ with diameter AH. Hence E is the circumcenter of triangle AMN. By the lemma, the bisector of $\angle BAC = \angle MAN$ intersects γ at the midpoint of the arc $\overset{\frown}{MN}$ not containing A. On the other hand, line OE also intersects γ at the midpoint of the arc $\overset{\frown}{MN}$, because M and N are symmetric respect to line OE. Hence the bisectors of $\angle BAC$ and $\angle MON$ meet at the midpoint of the arc $\overset{\frown}{MN}$. In particular, R lies on γ. By the following well-known and easy-to-prove geometry fact, R is the incenter of triangle OMN.

Let XYZ be a triangle with circumcenter O, circumcircle γ, and incenter I. Let X' be the second intersection of line XI with γ. Then X' is the circumcenter of triangle IYZ.

2. Find all polynomials $P(x)$ with real coefficients which satisfy the equality

$$P(a - b) + P(b - c) + P(c - a) = 2P(a + b + c)$$

for all triples (a, b, c) of real numbers such that $ab + bc + ca = 0$.

Note. We call a triple (a, b, c) of real numbers *good* if $ab + bc + ca = 0$. A simple but key observation is that if (a, b, c) is good, then so is (at, bt, ct), where t is an arbitrary real number.

First Solution. A polynomial $P(x)$ satisfies the condition of the problem if and only if $P(x) = c_1 x^2 + c_2 x^4$, where c_1 and c_2 are arbitrary real numbers.

We assume that $P(x)$ is a polynomial satisfying the conditions of the problem. Set

$$P(x) = p_n x^n + p_{n-1} x^{n-1} + \cdots + p_1 x + p_0 = \sum_{i=0}^{n} p_i x^i$$

for real numbers $p_0, p_1, \ldots, p_n$ with $p_n \neq 0$. For a good triple (at, bt, ct), we must have

$$\sum_{i=0}^{n} p_i t^i (a - b)^i + \sum_{i=0}^{n} p_i t^i (b - c)^i + \sum_{i=0}^{n} p_i t^i (c - a)^i$$

$$= 2 \sum_{i=0}^{n} p_i t^i (a + b + c)^i,$$

or

$$\sum_{i=0}^{n} p_i t^i [(a - b)^i + (b - c)^i + (c - a)^i - 2(a + b + c)^i] = 0. \quad (*)$$

Consider the polynomial

$$Q(x) = \sum_{i=0}^{n} q_i x^i = 0,$$

where $q_i = p_i[(a - b)^i + (b - c)^i + (c - a)^i - 2(a + b + c)^i]$. By equation $(*)$, we conclude that $Q(t) = 0$ for all real numbers t. Hence

$$q_i = p_i[(a - b)^i + (b - c)^i + (c - a)^i - 2(a + b + c)^i] = 0 \quad (*')$$

for all $0 \leq i \leq n$. In particular, if $i = 0$, then $p_i = 0$.

Because $ab + bc + ca = ab + c(a + b)$, setting $a + b = 1$ and $c = -ab$ leads to good triples. Hence we consider good triples of the form

$$(a, b, c) = [u, (1 - u), (u - 1)u].$$

Setting $u = 1$ and $u = 2$ in the above equation and substituting the resulting good triples in the equation $(*')$ provides valid but limited information. (Why?) Setting $u = 3$ gives $(a, b, c) = (3, -2, 6)$ and the equation $(*')$ becomes

$$p_i(5^i + (-8)^i + 3^i - 2 \cdot 7^i) = 0$$

for all $0 \le i \le n$. If i is odd, then

$$5^i + (-8)^i + 3^i - 2 \cdot 7^i = 5^i + 3^i - 8^i - 2 \cdot 7^i < 0,$$

and so $p_i = 0$. If i is an even integer greater than 4, then $\left(\frac{8}{7}\right)^i \ge \left(\frac{8}{7}\right)^6 = \frac{262144}{117649} > 2$, and so $5^i + (-8)^i + 3^i - 2 \cdot 7^i > 0$ and $p_i = 0$. (It is easy to check that $5^2 + 8^2 + 3^2 = 2 \cdot 7^2$ and $5^4 + 8^4 + 3^4 = 2 \cdot 7^4$.) Therefore, polynomials are of the form $P(x) = c_1 x^2 + c_2 x^4$ for some real numbers c_1 and c_2. We claim that any real c_1 and c_2 yield solutions of the problem. Note that if $P_1(x)$ and $P_2(x)$ are two polynomials satisfying the conditions of the problem, then $P(x) = c_1 P_1(x) + c_2 P_2(x)$, where c_1 and c_2 are arbitrary real numbers, is also a polynomial satisfying the conditions of the problem. Therefore, it suffices to show that $P_1(x) = x^2$ and $P_2(x) = x^4$ are solutions of the problem.

Let (a, b, c) be a good triple. We have

$$P_1(a - b) + P_1(b - c) + P_1(c - a)$$
$$= (a - b)^2 + (b - c)^2 + (c - a)^2$$
$$= 2(a^2 + b^2 + c^2) - 2(ab + bc + ca)$$
$$= 2(a^2 + b^2 + c^2) + 4(ab + bc + ca)$$
$$= 2(a + b + c)^2 = 2P_1(a + b + c),$$

implying that $P_1(x) = x^2$ satisfies the conditions of the problem. We show that $P_2(x) = x^4$ also satisfies the conditions of the problem; that is

$$f(a, b, c) = P_2(a-b) + P_2(b-c) + P_2(c-a) - 2P_2(a+b+c) = 0.$$

Because $(a + b + c)^2 = (a^2 + b^2 + c^2) - 2(ab + bc + ca) = (a^2 + b^2 + c^2)$, it follows that

$$2P_2(a + b + c) = 2(a + b + c)^4 = 2(a^2 + b^2 + c^2)^2$$
$$= 2(a^4 + b^4 + c^4) + 4(a^2b^2 + b^2c^2 + c^2a^2).$$

We have

$$P_2(a - b) + P_2(b - c) + P_2(c - a)$$
$$= (a - b)^4 + (b - c)^4 + (c - a)^4$$
$$= 2(a^4 + b^4 + c^4) - 4(a^3b + b^3c + c^3a)$$
$$+ 6(a^2b^2 + b^2c^2 + c^2a^2) - 4(ab^3 + bc^3 + ca^3).$$

Thus

$$f(a, b, c) = 2(a^2b^2 + b^2c^2 + c^2a^2)$$
$$- 4(a^3b + b^3c + c^3a + ab^3 + bc^3 + ca^3).$$

Note that

$$0 = (a^2 + b^2 + c^2)(ab + bc + ca)$$
$$= a^3b + b^3c + c^3a + ab^3 + bc^3 + ca^3 + (a^2bc + b^2ca + c^2ab).$$

It follows that

$$f(a, b, c) = 2[a^2b^2 + b^2c^2 + c^2a^2 + 2(a^2bc + b^2ca + c^2ab)]$$
$$= 2(ab + bc + ca)^2 = 0,$$

as desired.

Second Solution. Let $P(x)$ satisfy the given condition. If $a = b = 0$, then (a, b, c) is a good triple for all real numbers c. Hence

$$P(0 - 0) + P(0 - c) + P(c - 0) = 2P(0 + 0 + c),$$

or $P(c) = P(-c)$; that is, $P(x)$ is even. Setting $c = 0$, we also have $P(0) = 0$. Hence we can assume that

$$P(x) = p_{2n}x^{2n} + p_{2n-2}x^{2n-2} + \cdots + p_2x^2 = Q(x^2),$$

where $p_2, p_4, \ldots, p_{2n}$ are real numbers with $p_{2n} \neq 0$. The given condition becomes

$$Q((a-b)^2) + Q((b-c)^2) + Q((c-a)^2) = 2Q((a+b+c)^2) \quad (\dagger)$$

for all good triples (a, b, c). For a good triple (a, b, c), let $x = a - b$, $y = b - c$, and $z = c - a$. It is straightforward to check that

$$\begin{aligned}
(a+b+c)^2 &= \frac{x^2 + y^2 + z^2}{2} \\
&= \frac{x^2 + y^2 + (-x-y)^2}{2} \\
&= x^2 + xy + y^2.
\end{aligned}$$

The equation ($\dagger$) becomes

$$Q(x^2) + Q(y^2) + Q((x+y)^2) = 2Q(x^2 + xy + y^2).$$

Setting $x = y$ (that is, $x = y = a - b = b - c$ and $ab + bc + ca = 0$) in the last equation gives

$$2Q(x^2) + Q(4x^2) = 2Q(3x^2). \qquad (\dagger')$$

From $x = a - b = b - c$, we have $a = b + x$ and $c = b - x$. Substituting the last two equations into the equation $ab + bc + ca = 0$ gives $b^2 + bx + b^2 - bx + b^2 - x^2 = 0$, or $b = \frac{|x|}{\sqrt{3}}$. Hence for every real number x, there is a good triple

$$(a, b, c) = \left(\frac{|x|}{\sqrt{3}} + x, \frac{|x|}{\sqrt{3}}, \frac{|x|}{\sqrt{3}} - x \right)$$

such that $x = a - b = b - c$; that is, the equation ($\dagger'$) holds for all real numbers x. It follows that $2Q(x) + Q(4x) = 2Q(3x)$ holds for all nonnegative real numbers x, of which there are infinitely many. Because $Q(x)$ is a polynomial, we conclude that

$$2Q(x) + Q(4x) = 2Q(3x) \qquad (\ddagger)$$

for all real numbers x. Because $Q(x) = \sum_{i=1}^{n} p_i x^i$, the equation ($\ddagger$) implies that $2p_i + 4^i p_i = 2 \cdot 3^i p_i$, or $p_i(4^i + 2 - 2 \cdot 3^i) = 0$, for all $i = 1, 2, \ldots, n$. Because $\left(\frac{4}{3}\right)^i \geq \left(\frac{4}{3}\right)^3 > 2$ for $i \geq 3$, $p_i = 0$ for $i \geq 3$. Therefore, $Q(x) = c_1 x + c_2 x^2$, or $P(x) = Q(x^2) = c_1 x^2 + c_2 x^4$ for some real numbers c_1 and c_2. As shown in the first solution, we can verify that all such polynomials are the solutions of the problem.

Third Solution. (By Tony Zhang) As shown in the second solution, $P(x)$ is an even function and its constant term is 0; that is, $P(x) = x^2 f(x^2)$ for some polynomial $f(x)$, and we just need to show that $f(x)$ has degree at most 1. The given condition reads

$$(a-b)^2 f((a-b)^2) + (b-c)^2 f((b-c)^2) + (c-a)^2 f((c-a)^2)$$
$$= 2(a+b+c)^2 f((a+b+c)^2).$$

Plugging the good triple $(a, b, c) = [(1-\sqrt{3})b, b, (1+\sqrt{3})b]$ into the above equation gives

$$6b^2 f(3b^2) + 12b^2 f(12b^2) = 18b^2 f(9b^2),$$

which can be rewritten as

$$12b^2[f(12b^2) - f(9b^2)] = 6b^2[f(9b^2) - f(3b^2)]$$

or, for $b \neq 0$,

$$\frac{f(12b^2) - f(9b^2)}{3b^2} = \frac{f(9b^2) - f(3b^2)}{6b^2}$$

Consider points $A = (3b^2, f(3b^2))$, $B = (9b^2, f(9b^2))$, and $C = (12b^2, f(12b^2))$. Then the above equation says that lines AB and BC have the same slope. Note that if $P(x)$ is a solution, then so is $-P(x)$. Without loss of generality, we can assume the leading coefficient of $P(x)$ (and therefore $f(x)$) is nonnegative. If f is of degree 2 or higher, the second derivative of f is a polynomial with positive leading coefficient. Then there exist real numbers N such that for $x > N$, $f''(x) > 0$, and so $f(x)$ is convex. We choose b such that $3b^2 > N$. Then we have

$$\frac{f(12b^2) - f(9b^2)}{3b^2} > f'(9b^2)$$

$$> \frac{f(9b^2) - f(3b^2)}{6b^2};$$

that is, the slope of line line BC is greater than that of the line tangent to $f(x)$ at B, which is greater than that of AC, which is a contradiction. Therefore, f is at most linear; that is,

$$f(x) = c_1 + c_2 x \quad \text{and} \quad P(x) = x^2 f(x^2) = c_1 x^2 + c_2 x^4$$

for some real numbers c_1 and c_2. As shown in the first solution, we can verify that this is the complete solution set.

3. Define a *hook* to be a figure made up of six unit squares as shown in the diagram

or any of the figures obtained by applying rotations and reflections to this figure.

Determine all $m \times n$ rectangles that can be tiled with hooks so that

- the rectangle must be covered without gaps and without overlaps; and

- no part of a hook may cover area outside the rectangle.

Note. The rectangles that can be tiled with hooks are those with sides $\{3a, 4b\}$ or $\{c, 12d\}$ with $c \notin \{1, 2, 5\}$, where $a, b, c, d \in \mathbb{Z}^+$, and the order of the dimensions is not important.

We first show that these rectangles can be tiled. We can form a 3×4 rectangle from two hooks, so we can tile any $3a \times 4b$ rectangle. In particular, we can tile $3 \times 12d$ and $4 \times 12d$ rectangles, so by joining these along their long sides, we can tile a $c \times 12d$ rectangle for any $c \geq 6$.

Now we show that no other rectangles can be tiled with hooks. First, note that in any tiling of a rectangle by hooks, for any hook A its center square is covered by a unique hook B of the tiling, and the center square of B must be covered by A. Hence we can pair up the hooks in a tiling, and each pair of hooks will cover one of the (unrotated) shapes shown in Figure 3.1.

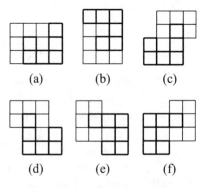

(a)   (b)   (c)

(d)   (e)   (f)

FIGURE 3.1.   2-hook "chunks"

The upshot is that given any tiling of a rectangle with hooks, it can be uniquely interpreted as a tiling with unrotated shapes of types (a) through (f), which we shall call "chunks." In particular, the area of the rectangle must be divisible by 12, because each shape has area 12. Also, it is then clear that no rectangle with a side of length 1, 2, or 5 can be tiled by these pieces. It remains to be shown that at least one side of the rectangle must be divisible by 4. Suppose we have a tiling of an $m \times n$ rectangle where neither m nor n is divisible by 4. Since $12 \mid mn$, m and n must both be even. We give five proofs that this is impossible.

First Solution. We observe that any chunk in the tiling has exactly one of the following two properties:

I: it consists of four adjacent columns, each containing three squares, and it has an even number of squares in each row, i.e., chunks (a), (e), (f);

II: it consists of four adjacent rows, each containing three squares, and it has an even number of squares in each column, i.e., chunks (b), (c), (d).

Refer to a chunk as "type I" or "type II" accordingly. Color the squares in every fourth row of the rectangle, as in Figure 3.2's example for 18×18.

A chunk of type I contains an even number of squares in each row, so it covers an even number of dark squares. A chunk of type II will intersect exactly one dark row, and it will contain three squares in that row, so it covers an odd number of dark squares. Since the rows of

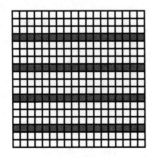

FIGURE 3.2.   Coloring for Solution 1

the rectangle have even length, the number of colored squares is even, so the number of chunks of type II is even. By a similar argument interchanging rows and columns, the number of chunks of type I is also even, so the total number of chunks is even. But then the total area mn must be divisible by $2 \times 12 = 24$, so at least one of m and n is divisible by 4, a contradiction.

Second Solution. Here is another way to show that $m, n \equiv 2 \pmod 4$ is impossible. Assume on the contrary that this is the case; imagine the rectangle divided into unit squares, with the rows and columns formed labeled $1, \ldots, m$ and $1, \ldots, n$ (from top to bottom and from left to right). Color the square in row i and column j if exactly one of i and j is divisible by 4, as shown in Figure 3.3's example for 18×18.

FIGURE 3.3. Coloring for Solution 2

Each of the six possible chunks covers 12 unit squares. Wherever they are, a straightforward verification shows that the chunk covers either 3 or 5 dark squares. Consequently, each chunk covers an *odd* number of dark squares.

Yet we can express $m = 4u + 2$, $n = 4v + 2$, so the total number of dark squares is

$$u(3v + 2) + v(3u + 2) = 2(3uv + u + v),$$

an even number. Hence the total number of chunks is even. As in the first solution, this forces mn to be divisible by $2 \times 12 = 24$, contradicting the fact that neither is divisible by 4.

Third Solution. This is a variant on the second solution; this time, write the digit 1 in the (i, j) square if exactly one of i and j is

			1				1				1				1		
			1				1				1				1		
			1				1				1				1		
1	1	1	2	1	1	1	2	1	1	1	2	1	1	1	2	1	1
			1				1				1				1		
			1				1				1				1		
			1				1				1				1		
1	1	1	2	1	1	1	2	1	1	1	2	1	1	1	2	1	1
			1				1				1				1		
			1				1				1				1		
			1				1				1				1		
1	1	1	2	1	1	1	2	1	1	1	2	1	1	1	2	1	1
			1				1				1				1		
			1				1				1				1		
			1				1				1				1		
1	1	1	2	1	1	1	2	1	1	1	2	1	1	1	2	1	1
			1				1				1				1		
			1				1				1				1		

FIGURE 3.4.   Numbering for Solution 3

divisible by 4, and the digit 2 if i and j are both divisible by 4. Do not write anything anywhere else (see Figure 3.4).

Since the number of squares in each row and column is even, the sum of all numbers written is even. Now, it is easy to check that a 3×4 rectangle always covers numbers with sum 3 or 7; and the other types of 12-square chunks always cover numbers with sum 5 or 7. Consequently, the total number of chunks is even, and we are done as before.

Fourth Solution. Here is a solution that uses two types of colorings to prove the impossibility of the $m, n \equiv 2 \pmod{4}$ case. The first proves that the total number of chunks of types (a) and (b) is even, and the last one proves that the total number of chunks of the other four types is also even, thus completing the proof as before. Consider the coloring in Figure 3.5 (shown for the example 18×18).

Clearly, the number of dark squares is equal to the number of white squares. Simple checking shows that of the six possible chunks, only type (a) can possibly yield a color imbalance. Furthermore, type (a) always yields a color imbalance, by exactly 4 units in one direction. Therefore, the number of type (a) chunks must be even. A similar coloring with respect to columns instead of rows proves that the number of type (b) chunks must be even as well.

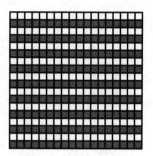

FIGURE 3.5.   First coloring for Solution 4

To prove that the number of chunks of types (c) through (f) is in fact even, consider the coloring in Figure 3.6.

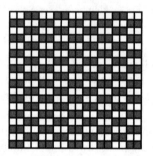

FIGURE 3.6.   Second coloring for Solution 4

Say that a chunk in the tiling of the $m \times n$ rectangle "starts on an even column" if the number of complete columns separating the chunk from the left side is even. Similarly define "start on an odd column," "start on an even row," etc. Simple verification shows that all chunks are balanced in color except for those of types (e) and (f) that start on even columns. In those special cases, the color imbalance is always 4 squares. It is clear that the coloring is balanced, so this proves that the number of types (e) and (f) that start on even columns is even. Now consider the coloring in Figure 3.7.

The same reasoning as before establishes that the number of types (e) and (f) that start on odd columns is even. Putting both results together, we find that the total number of chunks of types (e) and (f) is even. A similar argument involving columns instead of rows proves that the total number of chunks of types (c) and (d) is even.

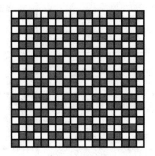

FIGURE 3.7.   Shift of second coloring for Solution 4

Putting everything together, we find that the total number of chunks of all types is even. This finishes the proof, as before.

Fifth Solution. Assume that $m, n \equiv 2 \pmod 4$. The second part of the fourth solution proves that the total number of chunks of types (c) through (f) is even. We can use the coloring in Figure 3.8 to prove that this number is in fact odd, yielding a contradiction.

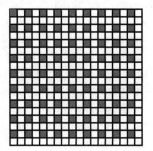

FIGURE 3.8.   Coloring for Solution 5

The number of dark squares is odd because $m, n \equiv 2 \pmod 4$; however, it is easy to check that chunks of types (a) and (b) always cover an even number of dark squares, while chunks of the other four types always cover exactly 3 dark squares. Therefore, by parity, the number of chunks of types (c) through (f) is indeed odd, giving a contradiction.

Note. This is the hardest problem on the exam. There are two traps.

First, observe that the set of dimensions of rectangles that can be tiled with hooks is the same as the set of dimensions of rectangles

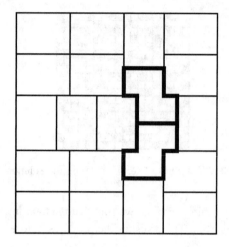

FIGURE 3.9.   Tiling using non-rectangular chunks (highlighted)

that can be tiled with 3×4 rectangles. Therefore, one might guess that there do not exist tilings using chunks of types (c) through (f). This, however, is false, as exemplified in Figure 3.9; there, a 15×16 rectangle is perfectly tiled using two such chunks.

Second, observe that four-coloring the board is a common way to prove that one dimension is divisible by four. Such coloring schemes do exist; however, the schemes and arguments are more complicated. Furthermore, as shown in many of the above solutions, more than one coloring scheme is needed to complete most proofs. The key to this problem is to cleverly use simple bichromatic schemes to analyze the orientations of the tiles.

4. Let n be an integer greater than or equal to 3, and let $t_1, t_2, \ldots, t_n$ be positive real numbers such that

$$n^2 + 1 > (t_1 + t_2 + \cdots + t_n)\left(\frac{1}{t_1} + \frac{1}{t_2} + \cdots + \frac{1}{t_n}\right).$$

Show that t_i, t_j, and t_k are side lengths of a triangle for all i, j, and k with $1 \leq i < j < k \leq n$.

Solution. We lose no generality by assuming that $t_1 \leq t_2 \leq \cdots \leq t_n$, so it suffices to show that $t_n < t_1 + t_2$. Expanding the right-hand side of the given inequality gives

$$n^2 + 1 > n + \sum_{1 \le i < j \le n} \left(\frac{t_i}{t_j} + \frac{t_j}{t_i} \right)$$

$$= n + t_n \left(\frac{1}{t_1} + \frac{1}{t_2} \right) + \frac{1}{t_n}(t_1 + t_2)$$

$$+ \sum_{\substack{1 \le i < j \le n \\ (i,j) \ne (1,n),(2,n)}} \left(\frac{t_i}{t_j} + \frac{t_j}{t_i} \right).$$

By the **AM-GM Inequality**, $\frac{t_i}{t_j} + \frac{t_j}{t_i} \ge 2$. There are $\binom{n}{2} = \frac{n(n-1)}{2}$ pairs of integers (i, j) with $1 \le i < j \le n$. It follows that

$$n^2 + 1 > n + t_n \left(\frac{1}{t_1} + \frac{1}{t_2} \right) + \frac{1}{t_n}(t_1 + t_2) + 2 \left[\binom{n}{2} - 2 \right]$$

or

$$t_n \left(\frac{1}{t_1} + \frac{1}{t_2} \right) + \frac{1}{t_n}(t_1 + t_2) - 5 < 0. \qquad (*)$$

By the AM-GM Inequality, $(t_1 + t_2) \left(\frac{1}{t_1} + \frac{1}{t_2} \right) = 2 + \frac{t_1}{t_2} + \frac{t_2}{t_1} \ge 4$, and so

$$\frac{4t_n}{t_1 + t_2} \le t_n \left(\frac{1}{t_1} + \frac{1}{t_2} \right).$$

Substituting the last inequality into inequality $(*)$ gives

$$\frac{4t_n}{t_1 + t_2} + \frac{1}{t_n}(t_1 + t_2) - 5 < 0.$$

Setting $\frac{t_1 + t_2}{t_n} = x$ in the last equality yields $\frac{4}{x} + x - 5 < 0$, or $0 > x^2 - 5x + 4 = (x - 1)(x - 4)$. It follows that $1 < x < 4$; that is, $t_n < t_1 + t_2 < 4t_n$, implying the desired result.

Note. One can determine the greatest number $f(n)$ such that, for positive real numbers $t_1, t_2, \ldots, t_n$, the inequality

$$f(n) > (t_1 + t_2 + \cdots + t_n) \left(\frac{1}{t_1} + \frac{1}{t_2} + \cdots + \frac{1}{t_n} \right)$$

implies that any t_i, t_j, t_k can be the side lengths of a triangle. The answer is

$$f(n) = (n + \sqrt{10} - 3)^2.$$

5. In a convex quadrilateral $ABCD$, diagonal BD bisects neither $\angle ABC$ nor $\angle CDA$. Point P lies inside quadrilateral $ABCD$ in such a way that

$$\angle PBC = \angle DBA \quad \text{and} \quad \angle PDC = \angle BDA.$$

Prove that quadrilateral $ABCD$ is cyclic if and only if $AP = CP$.

Note. We present five solutions. The first three of them are synthetic geometric proofs. Most students in the IMO had trouble dealing the uniqueness of geometric configurations (see the first two solutions) in the proof of the "if" part. The third solution only deals with the "if" part. It is rather independent of the geometric configurations. But on the other hand, its idea is remotely related to ideas in the proofs of the "only if" part in the first two solutions. The last two solutions (with one of them applying clever computations) do not rely on the configurations.

First note that the given conditions on the construction of point P are symmetric with respect to B and D, so we may assume without loss of generality that P lies inside triangle ACB (including the boundary).

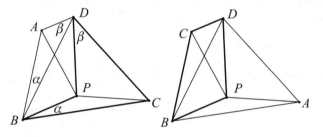

Next note that

$$\angle PBC = \angle DBA \quad \text{if and only if} \quad \angle PBA = \angle DBC,$$

and

$$\angle PDC = \angle BDA \quad \text{if and only if} \quad \angle PDA = \angle BDC.$$

Hence given conditions are symmetric respect to vertices A and C. Without loss of generality, we may assume that $\angle DBA$ is less than or $\angle DBC$. Then P lies inside triangle BDC; that is, we assume in all our solutions that we have the configuration shown on the left-hand side of the above figure. We set

$$\alpha = \angle PBC = \angle DBA \quad \text{and} \quad \beta = \angle PDC = \angle BDA.$$

First Solution.

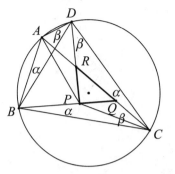

We first prove the "only if" part by assuming that $ABCD$ is cyclic. Let lines DP and BP meet segment AC at R and Q, respectively. Then $\angle RCD = \angle ACD = \angle ABD = \alpha$ and $\angle RCB = \angle ACB = \angle ADB = \beta$. Hence triangles ABD, QBC, and RCD are similar to each other. Because triangles QBC and RCD are similar, $\angle CQB = \angle DRC$, implying that $\angle QRP = \angle RQP$. It follows that triangle PQR is isosceles with $PQ = PR$. We conclude that

$$PQ = PR \quad \text{and} \quad \angle PQA = \angle PRC. \qquad (*)$$

Because triangles ABD and QBC are similar, we have $\frac{AB}{BD} = \frac{QB}{BC}$. Note that $\angle ABQ = \angle DBC$. It follows that triangle ABQ is similar to DBC. (That is, we obtain two pairs of triangles, namely $\{ABD, QBC\}$ and $\{ABC, DBC\}$, with **spiral similarity**.) Hence

$$\frac{AQ}{DC} = \frac{BQ}{BC}.$$

Because triangles DRC and CQB are similar, we have

$$\frac{RC}{DC} = \frac{QB}{CB}.$$

Combining the last two equations gives $AQ = RC$. Putting $AQ = RC$ together with the equations in $(*)$ shows that triangle APQ is congruent to triangle CPR, implying that $AP = CP$.

Now we prove the "if" part by assuming that $AP = CP$. Without loss of generality, we assume that P lies inside triangle ABC. Then D must lie on the ray BD such that $\angle DBA = \angle PBC$. Let ω denote the circumcircle of triangle ABC, and let D_1 be the (unique) point on $\overset{\frown}{AC}$ of ω such that $\angle D_1 BA = \angle PBC$. We will first prove that $ABCD_1$ and P satisfy the conditions of the problem, that is, $\angle AD_1B = \angle PD_1C$.

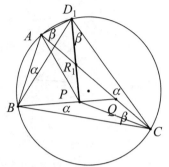

Extend segment BP through P to meet segment AC at Q. Set $\alpha = \angle D_1BA$ and $\beta = \angle BD_1A$. Then we have $\angle QCB = \angle ACB = \beta$ and $\angle QBC = \angle ACD_1 = \alpha$. Thus triangles AD_1B and QCB are similar, and so triangles ABQ and D_1BC are similar, implying that

$$\frac{AQ}{BQ} = \frac{D_1C}{BC}.$$

Construct the point R_1 on segment AC such that $CR_1 = AQ$. Substitute $CR_1 = AQ$ into the above equation:

$$\frac{CR_1}{BQ} = \frac{D_1C}{BC} \quad \text{or} \quad \frac{D_1C}{CR_1} = \frac{CB}{BQ}.$$

Combining the above equation with the fact that $\angle D_1CR_1 = \angle D_1CA = \angle CBQ$ shows that triangle D_1R_1C is similar to triangle CQB, implying that $\angle R_1D_1C = \angle QCB = \beta$ and $\angle D_1R_1C = \angle CQB$. Because $AP = PC$, $\angle QAP = \angle R_1CP$, and $AQ = CR_1$, triangles AQP and CR_1P are congruent. Hence

$$\angle QR_1P = \angle R_1QP = 180° - \angle CQB = 180° - \angle DR_1C,$$

and so D, R_1, and P are collinear. We conclude that

$$\angle PD_1C = \angle R_1D_1C = \beta = \angle BD_1A.$$

Hence cyclic quadrilateral $ABCD_1$ and point P satisfy the conditions of the problem.

To finish the solution of the problem, we have to show the uniqueness of point D with points A, B, C, and P fixed. (Also, without loss of generality, we may assume that P lies inside triangle ABC.) The proof is not trivial. We have the following lemma.

Lemma. *Let XYZ be a triangle, and let points S and T be such that lines XS and XT are symmetric with respect to the bisector of $\angle ZXY$, and that lines YS and YT are symmetric with respect to the bisector of $\angle XYZ$. Then lines ZS and ZT are symmetric with respect to the bisector of $\angle YZX$. Points S and T are said to be isogonal conjugates with respect to triangle XYZ.*

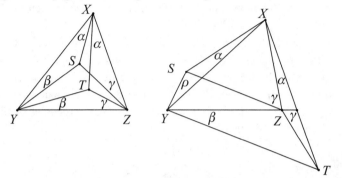

Proof: In order to consider both of the following configuration together, we use directed angles modulo $180°$. By the given condition, we have $\angle SXY = \angle ZXT = \alpha$ and $\angle XYS = \angle TYZ = \beta$, and it suffices to show that $\angle XZS = \angle TZY = \gamma$. Then $\angle ZXS = \angle TXY$ and $\angle SYZ = \angle XYT$. Applying the trigonometric form of **Ceva's Theorem** to triangles XYZ with S and then with T gives

$$\frac{\sin \angle SXY \, \sin \angle SYZ \, \sin \angle SZX}{\sin \angle ZXS \, \sin \angle XYS \, \sin \angle YZS} = 1,$$

$$\frac{\sin \angle TXY \, \sin \angle TYZ \, \sin \angle TZX}{\sin \angle ZXT \, \sin \angle XYT \, \sin \angle YZT} = 1.$$

Multiplying the last two equations together gives

$$\frac{\sin \angle SZX \, \sin \angle TZX}{\sin \angle YZS \, \sin \angle YZT} = 1.$$

Note that $\angle SZX = \angle YZT$ if and only if $\angle TZX = \angle YZS$. Considering the range of $\angle SZX$ and $\angle YZT$, we conclude that $\angle SZX = \angle YZT$, as desired.   ■

Let segments DP and AC meet at R. Note that A and C are isogonal conjugates respect to triangle BDP. By the lemma, $\angle APR = \angle CPQ$, which is fixed by points $A, B, C,$ and P. Therefore, point R is fixed, and indeed $R = R_1$, because $\angle R_1PA = \angle QPC$ by the congruency of triangle APQ and CR_1P. We conclude that $D_1 = D$, completing the proof.

Second Solution. We again assume first that $ABCD$ is cyclic. Extend segments BP and DP through P to meet the circumcircle of $ABCD$ at E and F, respectively. Then arc $\overset{\frown}{CE}$ (not containing A) is congruent to arc $\overset{\frown}{AD}$ (not containing C). Hence $ACDE$ is an isosceles trapezoid with $AC \parallel DE$. Likewise, $ABFC$ is an isosceles trapezoid with $AC \parallel BF$. Thus, $BF \parallel DE$, and so $BFED$ is an isosceles trapezoid with $BF \parallel DE$. Therefore, point P, which is the intersection of the diagonals BE and DF, lies on the perpendicular bisector ℓ of segments BF and DE. Line ℓ is also the perpendicular bisector of segment AC. We conclude that P lies on the perpendicular bisector of segment AC; that is, $AP = CP$.

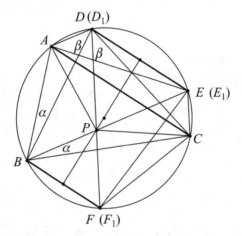

Now we prove the "if" part by assuming that $AP = CP$. We maintain the same notation as in the second part of the first solution. Let lines BP and D_1P meet ω again at E_1 and F_1. Since $\angle ABD_1 = \angle EBC$, quadrilateral AD_1E_1C is an isosceles trapezoid. Point P lies on its axis of symmetry because $AP = CP$, so P also lies on the perpendicular bisector of segment D_1E_1. Note also that P is the intersection of the diagonals BE_1 and D_1F_1 of cyclic quadrilateral $BD_1E_1F_1$. Hence $BD_1E_1F_1$ is an isosceles trapezoid with $BF_1 \parallel D_1E_1$ and $BD_1 = E_1F_1$. Thus ABF_1C is also an isosceles trapezoid with $AC \parallel BF_1$ and $AB = CF_1$, implying that $\angle AD_1B = \angle F_1D_1C$ as desired. This completes the proof of the existence of point D. The proof of the uniqueness of D is the same as shown in the first solution.

Third Solution. We present another approach proving the "if" part by assuming that $AP = CP$. Let lines BP and BC meet the circumcircle of triangle CDP at X and Y, respectively. (Note that it is possible that $YB > YC$. We use directed angles in our proof.)

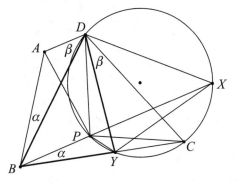

Because $DPYC$ is cyclic, $\angle PYB = \angle PDC = \angle ADB$. Hence triangles ABD and PBY are similar (spiral similarity), implying that triangles ABP and DBY are similar, implying that

$$\frac{AP}{DY} = \frac{BP}{BY}.$$

Because $CXPY$ is cyclic, $\angle BXY = \angle PXY = \angle PCY = \angle PCB$ Hence triangles BCP and BXY are similar, implying that

$$\frac{CP}{YX} = \frac{BP}{BY}.$$

Hence $\frac{AP}{DY} = \frac{CP}{YX}$, or $DY = XY$, because $AP = CP$. Because $DY = XY$, it follows that $\angle DCB = \angle DCY = \angle YPX = 180° - \angle BPY$, or $\angle DCB + \angle BPY = 180°$. Since triangle BAD is similar to triangle BPY, $\angle BAD = \angle BPY$. It follows that $\angle BCD + \angle BAD = 180°$, and hence $ABCD$ is cyclic.

Fourth Solution. Let ω' be the circumcircle of triangle BCD. Extend segments BP and DP through P to meet ω' again at E and F, respectively. Because $BFCE$ is cyclic, $\angle EFC = \angle EBC = \angle PBC = \angle ABD$. Likewise, $\angle FEC = \angle ADB$. Hence triangles ABD and CFE are similar, with ratio BD/EF. Because $BDFE$ is cyclic, triangles BDP and FEP are similar, with ratio BD/FE. It follows that quadrilaterals $ABPD$ and $CFPE$ are similar.

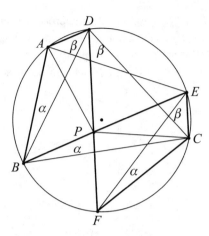

Therefore, $AP = CP$ if and only if quadrilaterals $ABPD$ and $CFPE$ are congruent, or $BD = EF$. Hence $AP = CP$ if and only if $\angle BFD = \angle EDF$. Because triangles ABD and CFE are similar, $\angle DAB = \angle ECF$. Because $DECF$ is cyclic, $AP = CP$ if and only if $\angle DAB + \angle BFD = \angle ECF + \angle EDF = 180°$; that is, if and only if $ABFD$ is cyclic. It follows that $AP = CP$ if and only if $ABFCED$ is cyclic.

Fifth Solution. (By Alexander Ivanov, observer with the Bulgarian team)

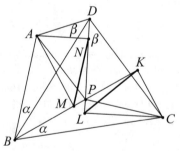

Let M, N, K, and L be the feet of the perpendiculars from A and C to lines BP and DP, as shown. Because $\angle AMP = \angle ANP = 90°$, points $A, M, P,$ and N lie on a circle and segment AP is a diameter of the circle. By the Extended Law of Sines,

$$MN = AP \sin \angle MAN = AP \sin(180° - \angle NPM)$$
$$= AP \sin \angle NPM.$$

Likewise, $KL = CP \sin \angle LPK$. Note that $\angle NPM = \angle LPK$. Hence $AP = CP$ if and only if $MN = KL$. Set $\angle A = \angle BAD$ and $\angle C = \angle DCB$. It suffices to show that $MN = KL$ if and only if $\angle A + \angle C = 180°$. Set $\gamma = \angle MBA$ and $\delta = \angle ADN$. Note that in triangles ABD and CDB, $\angle A = 180° - \alpha - \beta$ and $\angle C = 180° - \gamma - \delta$. Because $\gamma > \alpha$ and $\delta > \beta$, $\angle A > \angle C$. Hence $\angle A + \angle C = 180°$ if and only if $\sin \angle A = \sin \angle C$; that is, it suffices to show that $MN = KL$ if and only if $\sin \angle A = \sin \angle C$.

Set $\phi = \angle MAN$. Then $\angle CBD = \gamma$, $\angle BDC = \delta$, and $\angle KCL = \phi$. Applying the Law of Sines to triangle ABD gives

$$AB = \frac{BD \sin \beta}{\sin \angle A} \quad \text{and} \quad AD = \frac{BD \sin \alpha}{\sin \angle A}.$$

In right triangles AMB and AND, we have

$$AM = AB \sin \gamma = \frac{BD \sin \beta \sin \gamma}{\sin \angle A}$$

and

$$AN = AD \sin \delta = \frac{BD \sin \alpha \sin \delta}{\sin \angle A}.$$

Applying the **Law of Cosines** to triangle AMN gives

$$MN^2 = AM^2 + AN^2 - 2AM \cdot AN \cos \phi$$

$$= \frac{BD^2}{\sin^2 \angle A} \cdot f,$$

where $f = (\sin^2 \beta \sin^2 \gamma + \sin^2 \alpha \sin^2 \delta - 2 \sin \beta \sin \gamma \sin \alpha \sin \delta \cos \phi)$.

In exactly the same way, we can work on triangles BCD, BCK, CDL, and CLK to obtain

$$KL^2 = \frac{BD^2}{\sin^2 \angle C} \cdot f.$$

Hence $MN = KL$ if and only if $\sin \angle A = \sin \angle C$, as desired.

6. A positive integer is called *alternating* if among any two consecutive digits in its decimal representation, one is even and the other is odd. Find all positive integers n such that n has a multiple which is alternating.

Solution. The answers are those positive integers that are not divisible by 20. We call an integer n an *alternator* if it has a multiple which is alternating. Because any multiple of 20 ends with an even digit followed by 0, multiples of 20 are not alternating. Hence multiples of 20 are not alternators.

We consider separately the powers of 2 and the numbers of the form $2 \cdot 5^n$. In the sequel, $u^k \parallel a$ means that k is the largest integer such that u^k divides a, where u and a are positive integers, and $\overline{a_n \ldots a_1}$ denotes the integer with digits $a_n, a_{n-1}, \ldots, a_1$ (from left to right in that order).

Lemma 1. *Each power of 2 has a multiple which is alternating and has an even number of digits.*

Proof: It suffices to construct an infinite sequence $\{a_n\}_{n=1}^{\infty}$ of decimal digits such that

(a) $a_n \equiv n + 1 \pmod 2$;

(b) $2^{2n-1} \parallel \overline{a_{2n-1} \ldots a_1}$; and

(c) $2^{2n+1} \parallel \overline{a_{2n}a_{2n-1} \ldots a_1}$.

We construct this sequence inductively by starting with $a_1 = 2$ and $a_2 = 7$ and adding two digits in each step. Suppose the sequence is constructed up to a_{2n}. Set $a_{2n+1} = 4$. Then a_{2n+1} is even satisfying the condition (a).

Because $2^{2n+2} \parallel 4 \cdot 10^{2n}$ and $2^{2n+1} \parallel \overline{a_{2n}a_{2n-1} \ldots a_1}$ (by the induction hypothesis), we have

$$2^{2n+1} \parallel \overline{a_{2n}a_{2n-1} \ldots a_1} = 4 \cdot 10^{2n} + \overline{a_{2n}a_{2n-1} \ldots a_1},$$

establishing (b). Denote $\overline{a_{2n+1}a_{2n} \ldots a_1} = 2^{2n+1} \cdot A_n$. Then A_n is an odd integer. Now, a_{2n+2} must be odd and such that

$$2^{2n+3} \parallel \overline{a_{2n+2}a_{2n+1} \ldots a_1} = a_{2n+2} \cdot 10^{2n+1} + \overline{a_{2n+1}a_{2n} \ldots a_1}$$

$$= 2^{2n+1}(a_{2n+2}5^{2n+1} + A_n),$$

which holds whenever $5a_{2n+2} + A_n \equiv 4 \pmod 8$. The solutions of the last congruence are odd, since A_n is odd. In addition, because 5 is relatively prime to 8, a_{2n+2} can be chosen from $\{0, 1, \cdots, 7\}$, which is a complete residue class modulo 8. Thus our induction is complete. ∎

Lemma 2. *Each number of the form $2 \cdot 5^m$, where m is an arbitrary positive integer, has a multiple which is alternating and has an even number of digits.*

Proof: It suffices to construct an infinite sequence $\{b_n\}_{n=1}^{\infty}$ of decimal digits such that

(d) $b_n \equiv n + 1 \pmod 2$; and

(e) $2 \cdot 5^n \mid \overline{b_n b_{n-1} \ldots b_1}$.

We construct this sequence inductively by starting with $b_1 = 0$ and $b_2 = 5$ and adding one digit in each step. Suppose that $b_1, b_2, \ldots, b_n$ (with $n \geq 2$) have been constructed. Let $\overline{b_n b_{n-1} \ldots b_1} = 5^\ell B_n$, where B_n is a positive integer relatively prime to 5. By the induction hypothesis, $\ell \geq n$. The next digit b_{n+1} must be such that $b_{n+1} \equiv n+2 \pmod 2$ and $5^{n+1} \mid \overline{b_{n+1} b_n \ldots b_1}$. Note that

$$\overline{b_{n+1} b_n \ldots b_1} = b_{n+1} \cdot 2^n 5^n + 5^\ell B_n = 5^n (2^n b_{n+1} + 5^{\ell-n} B_n).$$

It suffices to find a digit b_{n+1} such that

$$b_{n+1} \equiv n + 2 \pmod 2 \quad \text{and} \quad 2^n b_{n+1} + 5^{\ell-n} B_n \equiv 0 \pmod 5.$$

Because 2 and 5 are relatively prime to each other, by the **Chinese Remainder Theorem**, such a b_{n+1} exists, moreover, a solution b_{n+1} can be chosen from the set $\{0, 1, 2, \ldots, 9\}$, which is a complete residue class modulo $2 \cdot 5 = 10$. Our induction is complete. ∎

We pass on to the general case of the numbers n written in the form of $n = 2^\alpha 5^\beta k$, where α and β are nonnegative integers and k is a positive integer relative prime to 10. We also assume that n is not a multiple of 20; that is, if $\beta \geq 1$, then $\alpha = 0$ or 1. We consider the following cases.

Case 1. In this case, we assume that $k = 1$ and $\beta = 0$; that is, $n = 2^\alpha$. By Lemma 1, n has a multiple M which is alternating and has an even number of digits.

Case 2. In this case, we assume that $k = 1$ and $\beta \geq 1$. Then $n = 5^\beta$ or $n = 2 \cdot 5^\beta$. By Lemma 2, n has a multiple M which is alternating and has an even number of digits.

Case 3. In this case, we assume that $k > 1$. By Cases 1 and 2, we conclude that $2^\alpha 5^\beta$ has a multiple M which is alternating and has an even number of digits. Because M has an even number of digits, it is not difficult to see that all integers $\overline{MM \ldots M}$ (copying M in an any number of times) are multiples of $2^\alpha 5^\beta$ that are alternating. We show that some of them are multiples of n, establishing the fact that n is an alternator when it is not a multiple of 20. Consider the sequence of numbers

$$m_1 = M, \quad m_2 = \overline{MM}, \quad m_3 = \overline{MMM}, \quad \ldots.$$

There are infinitely many numbers in the above sequence, so some two of them, say, m_i and m_j with $1 \le i < j$, are congruent to each other modulo k. Hence the number $m_j - m_i$ is divisible by k; that is,

$$0 \equiv m_j - m_i = \underbrace{MM \dots M}_{j-i \ M\text{'s}} \underbrace{00 \dots 0}_{2mi \ 0\text{'s}} \pmod{k},$$

where we assume that M has $2m$ digits for some positive integer m. Because k is relatively prime to 10, we conclude that

$$0 \equiv \underbrace{MM \dots M}_{j-i \ M\text{'s}} \pmod{k}.$$

Because $2^\alpha 5^\beta$ divides M and k and $2^\alpha 5^\beta$ are relatively prime, $n = 2^\alpha 5^\beta k$ divides $\underbrace{MM \dots M}_{j-i \ M\text{'s}}$, which is alternating, completing the proof.

Second Solution. (By Tiankai Liu) Note that all multiples of a non-alternator are bad. Every multiple of 20 ends in at least two even digits and is therefore not alternating, so all multiples of 20 are non-alternators. We claim that all other positive integers are alternators. Let n be positive integer not a multiple of 20. Note that all divisors of an alternator are alternators. We may assume that n is a even number. We establish the following lemma.

Lemma 3. If $n = 2^\ell$ or $2 \cdot 5^\ell$, for some positive integer ℓ, then there exists a multiple $X(n)$ of n such that $X(n)$ is alternating and $X(n)$ has n digits.

Proof: Set

$$M = \frac{10^{n+1} - 10}{99} = \underbrace{101010 \dots 10}_{n \text{ digits}}.$$

For every integer $k = 0, 1, \dots, n - 1$, there exists a sequence $e_0, e_1, \dots, e_k \in \{0, 2, 4, 6, 8\}$ such that

$$M + \sum_{i=0}^{k} e_i \cdot 10^i$$

is divisible by 2^{k+2} if n is of the form 2^ℓ, or by $2 \cdot 5^{k+1}$ if $n = 2 \cdot 5^\ell$. This is straightforwardly proved by induction on k. In particular, there exist $e_0, \ldots, e_{n-1} \in \{0, 2, 4, 6, 8\}$ such that

$$X(n) = M + \sum_{i=0}^{k} e_i \cdot 10^i$$

is divisible by n. This $X(n)$ is alternating and has n digits, as desired.
∎

Because n is even and not divisible by 20, we write n in the form
$n'm$, where $n' = 2^\ell$ or $2 \cdot 5^\ell$ and $\gcd(m, 10) = 1$. Let $c \geq n'$ be
an integer such that $10^c \equiv 1 \pmod{m}$. (Such a c exists because
$10^{\varphi(m)} \equiv 1 \pmod{m}$.) Let

$$M = \frac{10^{2mc+1} - 10}{99} \cdot 10^{n'} + X(n') = \underbrace{101010 \ldots 10}_{2mc \text{ digits}} X(n').$$

There exists $k \in \{0, 1, 2, \ldots, m - 1\}$ such that $M \equiv -2k \pmod{m}$.
Then

$$X(n) = M + \sum_{i=1}^{k} 2 \cdot 10^{ci}$$

is divisible by m. This $X(n)$ is also divisible by n' (as n' divides $10^{n'}$,
which divides 10^c) and is alternating. Thus n is an alternator.

Note. The USA IMO delegation pointed out that this problem is
connected to the following four known problems. However, most
team leaders believed that this problem contained sufficiently distinct
ingredients and chose it to be on the exam anyway.

- Prove that for every positive integer n there exists an n-digit
  number divisible by 5^n all of whose digits are odd.

- Prove that for every positive integer n there exists an n-digit
  number divisible by 2^n all of whose digits are nonzero.

- Determine all positive integers n such that n has a multiple whose
  digits are nonzero.

- [IMO 1994 Short List] A *wobbly number* is a positive integer
  whose digits are alternately nonzero and zero with the units digit
  being nonzero. Determine all positive integers which do not divide
  any wobbly numbers.

Problem Credits

USAMO

1. Titu Andreescu
2. Kiran Kedlaya and Lenny Ng
3. Ricky Liu
4. Melanie Wood
5. Titu Andreescu
6. Zuming Feng

Team Selection Test

1. Titu Andreescu
2. Po-Ru Loh
3. Ricky Liu
4. Titu Andreescu
5. Zuming Feng
6. Daniel Kane and Kiran Kedlaya

IMO

1. Romania
2. Republic of Korea
3. Estonia
4. Republic of Korea
5. Poland
6. Iran

5
Glossary

Addition-subtraction formulas   See **Trigonometric Identities**.

AM-GM Inequality   If $a_1, a_2, \ldots, a_n$ are n nonnegative real numbers, and if $m_1, m_2, \ldots, m_n$ are positive real numbers satisfying

$$m_1 + m_2 + \cdots + m_n = 1,$$

then

$$m_1 a_1 + m_2 a_2 + \cdots + m_n a_n \geq a_1^{m_1} a_2^{m_2} \cdots a_n^{m_n},$$

with equality if and only if $a_1 = a_2 = \cdots = a_n$.

Angle Bisector Theorem   Let ABC be a triangle, and let D be a point of side BC such that segment AD bisects $\angle BAC$. Then

$$\frac{AB}{AC} = \frac{BD}{DC}.$$

Cauchy-Schwarz Inequality   For any real numbers $a_1, a_2, \ldots, a_n$, and $b_1, b_2, \ldots, b_n$,

$$(a_1^2 + a_2^2 + \cdots + a_n^2)(b_1^2 + b_2^2 + \cdots + b_n^2)$$
$$\geq (a_1 b_1 + a_2 b_2 + \cdots + a_n b_n)^2,$$

with equality if and only if a_i and b_i are proportional with the same constant for all $i = 1, 2, \ldots, n$.

Ceva's Theorem and its trigonometric form   Let AD, BE, CF be three **cevians** of triangle ABC. The following are equivalent:

(i) AD, BE, CF are concurrent;

(ii) $\dfrac{AF}{FB} \cdot \dfrac{BD}{DC} \cdot \dfrac{CE}{EA} = 1;$

(iii) $\dfrac{\sin \angle ABE}{\sin \angle EBC} \cdot \dfrac{\sin \angle BCF}{\sin \angle FCA} \cdot \dfrac{\sin \angle CAD}{\sin \angle DAB} = 1.$

Cevian   A cevian of a triangle is any segment joining a vertex to a point on the opposite side.

Chinese Remainder Theorem   Let $\{m_k\}_{k=1}^{n}$ be a sequence of pairwise relatively prime positive integers, and let $\{r_k\}_{k=1}^{n}$ be a sequence of integers. Then the solution for x in the system:

$$x \equiv r_1 \pmod{m_1}$$

$$x \equiv r_2 \pmod{m_2}$$

$$\vdots$$

$$x \equiv r_n \pmod{m_n}$$

is precisely one of the residue classes modulo $m_1 m_2 \cdots m_n$.

Cyclic Sum   Let n be a positive integer. Given a function f of n variables, define the cyclic sum of variables $(x_1, x_2, \ldots, x_n)$ as

$$\sum_{\text{cyc}} f(x_1, x_2, \ldots, x_n) = f(x_1, x_2, \ldots, x_n) + f(x_2, x_3, \ldots, x_n, x_1)$$

$$+ \cdots + f(x_n, x_1, x_2, \ldots, x_{n-1}).$$

Dot Product   Let n be an integer greater then 1, and let $\mathbf{u} = [a_1, a_2, \ldots, a_n]$ and $\mathbf{v} = [b_1, b_2, \ldots, b_n]$ be two vectors. Define their *dot product* $\mathbf{u} \cdot \mathbf{v} = a_1 b_1 + a_2 b_2 + \cdots + a_n b_n$. It is easy to check that

(i) $\mathbf{v} \cdot \mathbf{v} = |\mathbf{v}|^2$, that is, the dot product of vector with itself is the square of the magnitude of $\mathbf{v}$ and $\mathbf{v} \cdot \mathbf{v} \geq 0$ with equality if and only if $\mathbf{v} = [0, 0, \ldots, 0]$;

(ii) $\mathbf{u} \cdot \mathbf{v} = \mathbf{v} \cdot \mathbf{u}$;

(iii) $\mathbf{u} \cdot (\mathbf{v} + \mathbf{w}) = \mathbf{u} \cdot \mathbf{v} + \mathbf{u} \cdot \mathbf{w}$, where $\mathbf{w}$ is a vector;

(iv) $(c\mathbf{u}) \cdot \mathbf{v} = c(\mathbf{u} \cdot \mathbf{v})$, where c is a scalar.

When vectors $\mathbf{u}$ and $\mathbf{v}$ are placed tail-by-tail at the origin O, let A and B be the tips of $\mathbf{u}$ and $\mathbf{v}$, respectively. Then $\overrightarrow{AB} = \mathbf{v} - \mathbf{u}$. Let $\angle AOB = \theta$.

Applying the **Law of Cosines** to triangle AOB yields

$$|\mathbf{v} - \mathbf{u}|^2 = AB^2 = OA^2 + OB^2 - 2OA \cdot OB \cos \theta$$
$$= |\mathbf{u}|^2 + |\mathbf{v}|^2 - 2|\mathbf{u}||\mathbf{v}| \cos \theta.$$

It follows that

$$(\mathbf{v} - \mathbf{u}) \cdot (\mathbf{v} - \mathbf{u}) = \mathbf{u} \cdot \mathbf{u} + \mathbf{v} \cdot \mathbf{v} - 2|\mathbf{u}||\mathbf{v}| \cos \theta,$$

or,

$$\cos \theta = \frac{\mathbf{u} \cdot \mathbf{v}}{|\mathbf{u}||\mathbf{v}|}.$$

Consequently, if $0 \le \theta \le 90°$, $\mathbf{u} \cdot \mathbf{v} \ge 0$. Considering the range of $\cos \theta$, we have provided a proof of the **Cauchy-Schwarz Inequality**.

Double-angle formulas   See **Trigonometric Identities**.

Euler (Totient) Function   Let n be a positive integer. The Euler function $\phi(n)$ is defined to be the number of integers between 1 and n that are relatively prime to n. The following are three fundamental properties of this function:

- $\phi(nm) = \phi(n)\phi(m)$ for positive integers m and n;

- if $n = p_1^{a_1} p_2^{a_2} \cdots p_k^{a_k}$ is a prime factorization of n (with distinct primes p_i), then

$$\phi(n) = n \left(1 - \frac{1}{p_1} \right) \left(1 - \frac{1}{p_2} \right) \cdots \left(1 - \frac{1}{p_k} \right);$$

- $\sum_{d \mid n} \phi(d) = n.$

Extended Law of Sines   In a triangle ABC with circumradius equal to R,

$$\frac{BC}{\sin A} = \frac{CA}{\sin B} = \frac{AB}{\sin C} = 2R.$$

Hölder's Inequality   Let $\{a_{ij}\}$ be a $m \times n$ matrix of numbers, and let $\{r_i\}_{i=1}^m$ be a sequence of positive real numbers such that $\sum_{i=1}^m r_i^{-1} = 1$. Then:

$$\sum_{j=1}^n \left| \prod_{i=1}^m a_{ij} \right| \le \prod_{i=1}^m \left(\sum_{j=1}^n |a_{ij}|^{r_i} \right)^{1/r_i}.$$

Jensen's Inequality   A *convex* function is a function $f(t)$ for which the second derivative is always nonnegative. This is equivalent to having the property that for any a, b in the domain, $f((a+b)/2) \leq (f(a) + f(b))/2$. Then given weights $\lambda_1, \lambda_2, \ldots, \lambda_n$ and real numbers $a_1, a_2, \ldots, a_n$:

$$f\left(\frac{\lambda_1 a_1 + \cdots + \lambda_n a_n}{\lambda_1 + \cdots + \lambda_n}\right) \leq \frac{\lambda_1 f(a_1) + \cdots + \lambda_n f(a_n)}{\lambda_1 + \cdots + \lambda_n}.$$

If $f(t)$ is *concave* (second derivative always nonpositive), then the direction of the inequality is reversed.

Kite   A kite is a quadrilateral that is symmetric with respect to at least one of its diagonals.

Law of Cosines   In a triangle ABC,

$$CA^2 = AB^2 + BC^2 - 2AB \cdot BC \cos \angle ABC,$$

and analogous equations hold for AB^2 and BC^2.

Menelaus's Theorem   Let ABC be a triangle, and let X, Y, and Z be points on lines AB, BC, and CA, respectively. Note that these are lines, not segments, so B may be between A and X, etc. Then, X, Y, and Z are collinear if and only if

$$\frac{AX}{XB} \frac{BY}{YC} \frac{CZ}{ZA} = -1,$$

where directed lengths are used (hence the -1).

Power of a Point Theorem   Given a fixed circle ω with center O and radius r, and a point P, we define the **power of P with respect to** ω to be $OP^2 - r^2$. Let ℓ be an arbitrary line containing P, and let its points of intersection with ω be X and Y. (If ℓ is tangent to ω, then we set both X and Y to the point of tangency.) Then, it turns out that (in terms of directed lengths) $OP^2 - r^2 = PX \cdot PY$. In particular, this means that if we also have another line m passing through P, then $PX \cdot PY = PA \cdot PB$, where A and B are the intersections of m with ω.

Product-to-sum formulas   See **Trigonometric Identities**.

Ptolemy's Theorem   Let $ABCD$ be a convex quadrilateral inscribed in a circle. Then $AB \cdot CD + BC \cdot DA = AC \cdot BD$, where AB denotes the length of side AB, etc.

Pythagorean Triple   This is an ordered triple of positive integers (a, b, c) with the property that $a^2 + b^2 = c^2$; that is, they can be the sides of a right triangle with hypotenuse c.

Radical Axis   Given two fixed circles, their radical axis is the set of points with equal **power** with respect to them. It is well-known to be perpendicular to the line connecting their centers, and if the circles intersect, it passes through the intersection points.

Spiral Similarity   See **Transformation**.

Transformation   A transformation of the plane is a mapping of the plane onto itself such that every point P is mapped into a unique image P' and every point Q' has a unique prototype (preimage, inverse image, counterimage) Q.

A **reflection across a line** (in the plane) is a transformation which takes every point in the plane into its mirror image, with the line as mirror. A **rotation** is a transformation when the entire plane is rotated about a fixed point in the plane.

A **similarity** is a transformation that preserves ratios of distances. If P' and Q' are the respective images of points P and Q under a similarity $\mathbf{T}$, then the ratio $P'Q'/PQ$ depends only on $\mathbf{T}$. This ratio is the **similitude** of $\mathbf{T}$. A **dilation** is a direction-preserving similarity, i.e., a similarity that takes each line into a parallel line.

The **product $\mathbf{T_2 T_1}$ of two transformations** is the transformation defined by $\mathbf{T_2 T_1} = T_2 \circ T_1$, where $\circ$ denotes function composition. A **spiral similarity** is the product of a rotation and a dilation, or vice versa.

Triangle Inequality   Let $z = a + bi$ be a complex number. Define the absolute value of z to be

$$|z| = \sqrt{a^2 + b^2}.$$

Let α and β be two complex numbers. The inequality

$$|\alpha + \beta| \leq |\alpha| + |\beta|$$

is called the triangle inequality.

Let $\alpha = \alpha_1 + \alpha_2 i$ and $\beta = \beta_1 + \beta_2 i$, where $\alpha_1, \alpha_2, \beta_1, \beta_2$ are real numbers. Then $\alpha + \beta = (\alpha_1 + \beta_1) + (\alpha_2 + \beta_2)i$. Vectors $\mathbf{u} = [\alpha_1, \alpha_2]$, $\mathbf{v} = [\beta_1, \beta_2]$, and $\mathbf{w} = [\alpha_1 + \beta_1, \alpha_2 + \beta_2]$ form a triangle with sides lengths $|\alpha|$, $|\beta|$, and

$|\alpha + \beta|$. The triangle inequality restates the fact the the length of any side of a triangle is less than the sum of the lengths of the other two sides.

Trigonometric Identities

$$\sin^2 x + \cos^2 x = 1,$$

$$\tan x = \frac{\sin x}{\cos x}, \quad \cot x = \frac{1}{\tan x},$$

$$\sin(-x) = -\sin x, \quad \cos(-x) = \cos x,$$

$$\tan(-x) = -\tan x, \quad \cot(-x) = -\cot x,$$

$$\sin(90° \pm x) = \cos x, \quad \cos(90° \pm x) = \mp \sin x,$$

$$\tan(90° \pm x) = \mp \cot x, \quad \cot(90° \pm x) = \mp \tan x,$$

$$\sin(180° \pm x) = \mp \sin x, \quad \cos(180° \pm x) = -\cos x,$$

$$\tan(180° \pm x) = \pm \tan x, \quad \cot(180° \pm x) = \pm \cot x.$$

Addition and subtraction formulas:

$$\sin(a \pm b) = \sin a \cos b \pm \cos a \sin b,$$

$$\cos(a \pm b) = \cos a \cos b \mp \sin a \sin b,$$

$$\tan(a \pm b) = \frac{\tan a \pm \tan b}{1 \mp \tan a \tan b}.$$

Double-angle formulas:

$$\sin 2a = 2 \sin a \cos a,$$

$$\cos 2a = 2 \cos^2 a - 1 = 1 - 2 \sin^2 a = \cos^2 \alpha - \sin^2 \alpha,$$

$$\tan 2a = \frac{2 \tan a}{1 - \tan^2 a}.$$

Triple-angle formulas:

$$\sin 3a = 3 \sin a - 4 \sin^3 a = (3 - 4 \sin^2 a) \sin a = (4 \cos^2 a - 1) \sin a,$$

$$\cos 3a = 4 \cos^3 a - 3 \cos a = (4 \cos^2 -3) \cos a = (1 - 4 \sin^2 a) \cos a,$$

$$\tan 3a = \frac{3 \tan a - \tan^3 a}{1 - 3 \tan^2 a}.$$

Half-angle formulas:

$$\sin^2 \frac{a}{2} = \frac{1 - \cos a}{2},$$

$$\cos^2 \frac{a}{2} = \frac{1 + \cos a}{2}.$$

Sum-to-product formulas:

$$\sin a + \sin b = 2 \sin \frac{a+b}{2} \cos \frac{a-b}{2},$$

$$\cos a + \cos b = 2 \cos \frac{a+b}{2} \cos \frac{a-b}{2},$$

$$\tan a + \tan b = \frac{\sin(a+b)}{\cos a \cos b}.$$

Difference-to-product formulas:

$$\sin a - \sin b = 2 \sin \frac{a-b}{2} \cos \frac{a+b}{2},$$

$$\cos a - \cos b = -2 \sin \frac{a-b}{2} \sin \frac{a+b}{2},$$

$$\tan a - \tan b = \frac{\sin(a-b)}{\cos a \cos b}.$$

Product-to-sum formulas:

$$2 \sin a \cos b = \sin(a+b) + \sin(a-b),$$

$$2 \cos a \cos b = \cos(a+b) + \cos(a-b),$$

$$2 \sin a \sin b = -\cos(a+b) + \cos(a-b).$$

Expansion formulas

$$\sin n\alpha = \binom{n}{1} \cos^{n-1} \alpha \sin \alpha - \binom{n}{3} \cos^{n-3} \alpha \sin^3 \alpha$$

$$+ \binom{n}{5} \cos^{n-5} \alpha \sin^5 \alpha - \cdots,$$

$$\cos n\alpha = \binom{n}{0} \cos^n \alpha - \binom{n}{2} \cos^{n-2} \alpha \sin^2 \alpha$$

$$+ \binom{n}{4} \cos^{n-4} \alpha \sin^4 \alpha - \cdots.$$

6
Further Reading

1. Andreescu, T.; Feng, Z., *USA and International Mathematical Olympiads 2003* , Mathematical Association of America, 2004.

2. ——, *USA and International Mathematical Olympiads 2002* , Mathematical Association of America, 2003.

3. ——, *USA and International Mathematical Olympiads 2001* , Mathematical Association of America, 2002.

4. ——, *USA and International Mathematical Olympiads 2000* , Mathematical Association of America, 2001.

5. Andreescu, T.; Feng, Z.; Lee, G.; Loh, P., *Mathematical Olympiads: Problems and Solutions from around the World, 2001–2002*, Mathematical Association of America, 2005.

6. ——, *Mathematical Olympiads: Problems and Solutions from around the World, 2000–2001*, Mathematical Association of America, 2003.

7. Andreescu, T.; Feng, Z., *Mathematical Olympiads: Problems and Solutions from around the World, 1999–2000*, Mathematical Association of America, 2002.

8. ——, *Mathematical Olympiads: Problems and Solutions from around the World, 1998–1999*, Mathematical Association of America, 2000.

9. Andreescu, T.; Kedlaya, K., *Mathematical Contests 1997–1998: Olympiad Problems from around the World, with Solutions*, American Mathematics Competitions, 1999.

10. ——, *Mathematical Contests 1996–1997: Olympiad Problems from around the World, with Solutions*, American Mathematics Competitions, 1998.

11. Andreescu, T.; Kedlaya, K.; Zeitz, P., *Mathematical Contests 1995–1996: Olympiad Problems from around the World, with Solutions*, American Mathematics Competitions, 1997.

12. Andreescu, T.; Feng, Z., *103 Trigonometry Problems from the Training of the USA IMO Team*, Birkhäuser, 2004.

13. ——, *102 Combinatorial Problems from the Training of the USA IMO Team*, Birkhäuser, 2002.

14. ——, *101 Problems in Algebra from the Training of the USA IMO Team*, Australian Mathematics Trust, 2001.

15. ——, *A Path to Combinatorics for Undergraduate Students: Counting Strategies*, Birkhäuser, 2003.

16. Andreescu, T.; Enescu, B., *Mathematical Olympiad Treasures*, Birkhäuser, 2003.

17. Andreescu, T.; Gelca, R., *Mathematical Olympiad Challenges*, Birkhäuser, 2000.

18. Andreescu, T.; Andrica, D., *360 Problems for Mathematical Contests*, GIL Publishing House, 2003.

19. ——, *An Introduction to Diophantine Equations*, GIL Publishing House, 2002.

20. ——, *Complex Numbers from A to . . . Z*, Birkhäuser, 2005.

21. Barbeau, E., *Polynomials*, Springer-Verlag, 1989.

22. Beckenbach, E. F.; Bellman, R., *An Introduction to Inequalities*, New Mathematical Library, Vol. 3, Mathematical Association of America, 1961.

23. Bollobas, B., *Graph Theory, An Introductory Course*, Springer-Verlag, 1979.

24. Chinn, W. G.; Steenrod, N. E., *First Concepts of Topology*, New Mathematical Library, Vol. 27, Mathematical Association of America, 1966.

25. Cofman, J., *What to Solve?*, Oxford Science Publications, 1990.

26. Coxeter, H. S. M.; Greitzer, S. L., *Geometry Revisited*, New Mathematical Library, Vol. 19, Mathematical Association of America, 1967.

27. Coxeter, H. S. M., *Non-Euclidean Geometry*, The Mathematical Association of America, 1998.

28. Doob, M., *The Canadian Mathematical Olympiad 1969–1993*, University of Toronto Press, 1993.

29. Engel, A., *Problem-Solving Strategies*, Problem Books in Mathematics, Springer, 1998.

30. Fomin, D.; Kirichenko, A., *Leningrad Mathematical Olympiads 1987–1991*, MathPro Press, 1994.

31. Fomin, D.; Genkin, S.; Itenberg, I., *Mathematical Circles*, American Mathematical Society, 1996.

32. Graham, R. L.; Knuth, D. E.; Patashnik, O., *Concrete Mathematics*, Addison-Wesley, 1989.

33. Gillman, R., *A Friendly Mathematics Competition*, The Mathematical Association of America, 2003.

34. Greitzer, S. L., *International Mathematical Olympiads, 1959–1977*, New Mathematical Library, Vol. 27, Mathematical Association of America, 1978.

35. Grossman, I.; Magnus, W., *Groups and Their Graphs*, New Mathematical Library, Vol. 14, Mathematical Association of America, 1964.

36. Holton, D., *Let's Solve Some Math Problems*, A Canadian Mathematics Competition Publication, 1993.

37. Ireland, K.; Rosen, M., *A Classical Introduction to Modern Number Theory*, Springer-Verlag, 1982.

38. Kazarinoff, N. D., *Geometric Inequalities*, New Mathematical Library, Vol. 4, Mathematical Association of America, 1961.

39. Kedlaya, K; Poonen, B.; Vakil, R., *The William Lowell Putnam Mathematical Competition 1985–2000*, The Mathematical Association of America, 2002.

40. Klamkin, M., *International Mathematical Olympiads, 1978–1985*, New Mathematical Library, Vol. 31, Mathematical Association of America, 1986.

41. ——, *USA Mathematical Olympiads, 1972–1986*, New Mathematical Library, Vol. 33, Mathematical Association of America, 1988.

42. Klee, V.; Wagon, S, *Old and New Unsolved Problems in Plane Geometry and Number Theory*, The Mathematical Association of America, 1991.

43. Kürschák, J., *Hungarian Problem Book, volumes I & II*, New Mathematical Library, Vols. 11 & 12, Mathematical Association of America, 1967.

44. Kuczma, M., *144 Problems of the Austrian–Polish Mathematics Competition 1978–1993*, The Academic Distribution Center, 1994.

45. ——, *International Mathematical Olympiads 1986–1999*, Mathematical Association of America, 2003.

46. Landau, E., *Elementary Number Theory*, Chelsea Publishing Company, New York, 1966.

47. Larson, L. C., *Problem-Solving Through Problems*, Springer-Verlag, 1983.

48. Lausch, H. *The Asian Pacific Mathematics Olympiad 1989–1993*, Australian Mathematics Trust, 1994.

49. Leveque, W. J., *Topics in Number Theory, Volume 1*, Addison Wesley, New York, 1956.

50. Liu, A., *Chinese Mathematics Competitions and Olympiads 1981–1993*, Australian Mathematics Trust, 1998.

51. ——, *Hungarian Problem Book III*, New Mathematical Library, Vol. 42, Mathematical Association of America, 2001.

52. Lozansky, E.; Rousseau, C. *Winning Solutions*, Springer, 1996.

53. Mitrinovic, D. S.; Pecaric, J. E.; Volonec, V. *Recent Advances in Geometric Inequalities*, Kluwer Academic Publisher, 1989.

54. Mordell, L. J., *Diophantine Equations*, Academic Press, London and New York, 1969.

55. Ore, O., *Graphs and Their Uses*, New Mathematical Library, MAA, 1963.

56. ——, *Invitation to Number Theory*, New Mathematical Library, MAA, 1967.

57. Savchev, S.; Andreescu, T. *Mathematical Miniatures*, Anneli Lax New Mathematical Library, Vol. 43, Mathematical Association of America, 2002.

58. Sharygin, I. F., *Problems in Plane Geometry*, Mir, Moscow, 1988.

59. ——, *Problems in Solid Geometry*, Mir, Moscow, 1986.

60. Shklarsky, D. O; Chentzov, N. N; Yaglom, I. M., *The USSR Olympiad Problem Book*, Freeman, 1962.

61. Slinko, A., *USSR Mathematical Olympiads 1989–1992*, Australian Mathematics Trust, 1997.

62. Sierpinski, W., *Elementary Theory of Numbers*, Hafner, New York, 1964.

63. Soifer, A., *Colorado Mathematical Olympiad: The first ten years*, Center for excellence in mathematics education, 1994.

64. Szekely, G. J., *Contests in Higher Mathematics*, Springer-Verlag, 1996.

65. Stanley, R. P., *Enumerative Combinatorics*, Cambridge University Press, 1997.

66. Tabachnikov, S. *Kvant Selecta: Algebra and Analysis I*, American Mathematical Society, 1991.

67. ——, *Kvant Selecta: Algebra and Analysis II*, American Mathematical Society, 1991.

68. ——, *Kvant Selecta: Combinatorics I*, American Mathematical Society, 2000.

69. Taylor, P. J., *Tournament of Towns 1980–1984*, Australian Mathematics Trust, 1993.

70. ——, *Tournament of Towns 1984–1989*, Australian Mathematics Trust, 1992.

71. ——, *Tournament of Towns 1989–1993*, Australian Mathematics Trust, 1994.

72. Taylor, P. J.; Storozhev, A., *Tournament of Towns 1993–1997*, Australian Mathematics Trust, 1998.

73. Tomescu, I., *Problems in Combinatorics and Graph Theory*, Wiley, 1985.

74. Vanden Eynden, C., *Elementary Number Theory*, McGraw-Hill, 1987.

75. Vaderlind, P.; Guy, R.; Larson, L., *The Inquisitive Problem Solver*, The Mathematical Association of America, 2002.

76. Wilf, H. S., *Generatingfunctionology*, Academic Press, 1994.

77. Wilson, R., *Introduction to graph theory*, Academic Press, 1972.

78. Yaglom, I. M., *Geometric Transformations*, New Mathematical Library, Vol. 8, MAA, 1962.

79. Yaglom, I. M., *Geometric Transformations II*, New Mathematical Library, Vol. 21, MAA, 1968.

80. Yaglom, I. M., *Geometric Transformations III*, New Mathematical Library, Vol. 24, MAA, 1973.

81. Zeitz, P., *The Art and Craft of Problem Solving*, John Wiley & Sons, 1999.

7
Appendix

1 2004 Olympiad Results

Tiankai Liu, with a perfect score, was the winner of the Samuel Greitzer-Murray Klamkin award, given to the top scorer(s) on the USAMO. Tiankai is the winner of this award for the third year in a row. Tiankai Liu, Oleg Golberg, and Tony Zhang placed first, second, and third, respectively, on the USAMO. They were awarded college scholarships of $15000, $10000, and $5000, respectively, by the Akamai Foundation. The Clay Mathematics Institute (CMI) award, for a solution of outstanding elegance, and carrying a $3000 cash prize, was presented to Matt Ince for his solution to USAMO Problem 2, presented as the third solution to the problem in this book.

The top twelve students on the 2004 USAMO were (in alphabetical order):

Jae Bae	Academy of Advancement in Science and Technology	Hackensack, NJ
Jongmin Baek	Cupertino High School	Cupertino, CA
Oleg Golberg	Homeschooled	Exeter, NH
Matt Ince	St. Louis Family Church Learning Center	Chesterfield, MO
Janos Kramar	University of Toronto Schools	Toronto, ON
Tiankai Liu	Phillips Exeter Academy	Exeter, NH
Alison Miller	Home Educators Enrichment Group	Niskayuna, NY
Aaron Pixton	Vestal Senior High School	Vestal, NY
Brian Rice	Southwest Virginia Governor's School	Dublin, VA

Jacob Tsimerman	University of Toronto Schools	Toronto, ON
Ameya Velingker	Parkland High School	Allentown, PA
Tony Zhang	Phillips Exeter Academy	Exeter, NH

The USA team members were chosen according to their combined performance on the 33rd annual USAMO, and the Team Selection Test that took place at the Mathematical Olympiad Summer Program (MOSP), held at the University of Nebraska-Lincoln, June 13–July 3, 2004. Members of the USA team at the 2004 IMO (Athens, Greece) were Oleg Golberg, Matt Ince, Tiankai Liu, Alison Miller, Aaron Pixton, and Tony Zhang. Zuming Feng (Phillips Exeter Academy) and Po-Shen Loh (California Institute of Technology) served as team leader and deputy leader, respectively. The team was also accompanied by Reid Barton (Massachusetts Institute of Technology), Zvezdelina Stankova (Mills College), and Steven Dunbar (University of Nebraska-Lincoln), as observers of the team leader and deputy leader, respectively. During the competition, Professor Edward Witten (Institute for Advanced Study in Princeton, Fields medal winner) visited the team and congratulated the team's performance.

At the 2004 IMO, gold medals were awarded to students scoring between 32 and 42 points, silver medals to students scoring between 24 and 31 points, and bronze medals to students scoring between 16 and 23 points. There were 45 gold medalists, 78 silver medalists, and 120 bronze medalists. There were 4 perfect papers (Tsimerman from Canada, Rácz from Hungary, and Badzyan and Dubashinskiy from Russian Federation) on this difficult exam. Golberg's score of 40 tied for 7th place overall. Miller became the first female gold medalist from our country. The team's individual performances were as follows:

Golberg	GOLD Medallist	Liu	GOLD Medallist
Ince	SILVER Medallist	Pixton	GOLD Medallist
Miller	GOLD Medallist	Zhang	GOLD Medallist

The total team scores on the first five problems were all even for the top three teams, namely, China, USA, and Russian Federation. Therefore, the total team scores (China 42, USA 34, and Russia 27) on the sixth problem become the deciding factor of the team rankings. In terms of total score (out of a maximum of 252), the highest ranking of the 85 participating teams were as follows:

China	220	Hungary	187
USA	212	Japan	182
Russia	205	Iran	177
Vietnam	196	Romania	176
Bulgaria	194	Ukraine	174
Taiwan	190	Korea	166

The 2004 USAMO was prepared by Titu Andreescu (Chair), Steven Dunbar, Zuming Feng, Kiran Kedlaya, Alexander Soifer, Richard Stong, Zoran Sunik, Zvezdelina Stankova, and David Wells. The Team Selection Test was prepared by Titu Andreescu and Zuming Feng. The MOSP was held at the University of Nebraska-Lincoln. Because of a generous grant from the Akamai Foundation, the 2004 MOSP expanded from the usual 24-30 students to 54. An adequate number of instructors and assistants were appointed. Zuming Feng (Academic Director), Titu Andreescu, Chris Jeuell, Qin Jing, Po-Shen Loh, Alex Saltman, and Zvezdelina Stankova served as instructors, assisted by Reid Barton as junior instructor, and Mark Lipson, Ricky Liu, Po-Ru Loh, Gregory Price, and Inna Zakharevich as graders. Steven Dunbar (MOSP Director) and Kiran Kedlaya served as guest instructors.

For more information about the USAMO or the MOSP, contact Steven Dunbar at `sdunbar@math.unl.edu`.

2   2003 Olympiad Results

Tiankai Liu and Po-Ru Loh, both with perfect scores, were the winners of the Samuel Greitzer-Murray Klamkin award, given to the top scorer(s) on the USAMO. Mark Lipson placed third on the USAMO. They were awarded college scholarships of $5000, $5000, $2000, respectively, by the Akamai Foundation. The Clay Mathematics Institute (CMI) award, for a solution of outstanding elegance, and carrying a $3000 cash prize, was presented to Tiankai Liu for his solution to USAMO Problem 6. Two additional CMI awards, carrying a $1000 cash prize each, were presented to Anders Kaseorg and Matthew Tang for their solutions to USAMO Problem 5.

The top twelve students on the 2003 USAMO were (in alphabetical order):

Boris Alexeev	Cedar Shoals High School	Athens, GA
Jae Bae	Academy of Advancement in Science and Technology	Hackensack, NJ

Daniel Kane	West High School	Madison, WI
Anders Kaseorg	Charlotte Home Educators Association	Charlotte, NC
Mark Lipson	Lexington High School	Lexington, MA
Tiankai Liu	Phillips Exeter Academy	Exeter, NH
Po-Ru Loh	James Madison Memorial High School	Madison, WI
Po-Ling Loh	James Madison Memorial High School	Madison, WI
Aaron Pixton	Vestal Senior High School	Vestal, NY
Kwokfung Tang	Phillips Exeter Academy	Exeter, NH
Tony Zhang	Phillips Exeter Academy	Exeter, NH
Yan Zhang	Thomas Jefferson High School of Science and Technology	Alexandria, VA

The USA team members were chosen according to their combined performance on the 32nd annual USAMO and the Team Selection Test that took place at the Mathematical Olympiad Summer Program (MOSP) held at the University of Nebraska-Lincoln, June 15 - July 5, 2003. Members of the USA team at the 2003 IMO (Tokyo, Japan) were Daniel Kane, Anders Kaseorg, Mark Lipson, Po-Ru Loh, Aaron Pixton, and Yan Zhang. Zuming Feng (Phillips Exeter Academy) and Gregory Galperin (Eastern Illinois University) served as team leader and deputy leader, respectively. The team was also accompanied by Melanie Wood (Princeton University) and Steven Dunbar (University of Nebraska-Lincoln), as observers of the team leader and deputy leader, respectively.

At the 2003 IMO, gold medals were awarded to students scoring between 29 and 42 points, silver medals to students scoring between 19 and 28 points, and bronze medals to students scoring between 13 and 18 points. There were 37 gold medalists, 69 silver medalists, and 104 bronze medalists. There were three perfect papers (Fu from China, Le and Nguyen from Vietnam) on this very difficult exam. Loh's 36 tied for 12th place overall. The team's individual performances were as follows:

Kane	GOLD Medallist	Loh	GOLD Medallist
Kaseorg	GOLD Medallist	Pixton	GOLD Medallist
Lipson	SILVER Medallist	Y. Zhang	SILVER Medallist

In terms of total score (out of a maximum of 252), the highest ranking of the 82 participating teams were as follows:

Bulgaria	227	Romania	143
China	211	Turkey	133
USA	188	Japan	131
Vietnam	172	Hungary	128
Russia	167	United Kingdom	128
Korea	157	Canada	119
		Kazakhstan	119

The 2003 USAMO was prepared by Titu Andreescu (Chair), Zuming Feng, Kiran Kedlaya, and Richard Stong. The Team Selection Test was prepared by Titu Andreescu and Zuming Feng. The MOSP was held at the University of Nebraska-Lincoln. Zuming Feng (Academic Director), Gregory Galperin, and Melanie Wood served as instructors, assisted by Po-Shen Loh and Reid Barton as junior instructors, and Ian Le and Ricky Liu as graders. Kiran Kedlaya served as guest instructor.

3 2002 Olympiad Results

Daniel Kane, Ricky Liu, Tiankai Liu, Po-Ru Loh, and Inna Zakharevich, all with perfect scores, tied for first on the USAMO. They shared college scholarships of $30000 provided by the Akamai Foundation. The Clay Mathematics Institute (CMI) award, for a solution of outstanding elegance, and carrying a $1000 cash prize, was presented to Michael Hamburg, for the second year in a row, for his solution to USAMO Problem 6.

The top twelve students on the 2002 USAMO were (in alphabetical order):

Steve Byrnes	Roxbury Latin School	West Roxbury, MA
Michael Hamburg	Saint Joseph High School	South Bend, IN
Neil Herriot	Palo Alto High School	Palo Alto, CA
Daniel Kane	West High School	Madison, WI
Anders Kaseorg	Charlotte Home Educators Association	Charlotte, NC
Ricky Liu	Newton South High School	Newton, MA
Tiankai Liu	Phillips Exeter Academy	Exeter, NH
Po-Ling Loh	James Madison Memorial High School	Madison, WI
Alison Miller	Home Educators Enrichment Group	Niskayuna, NY

Gregory Price	Thomas Jefferson High School of Science and Technology	Alexandria, VA
Tong-ke Xue	Hamilton High School	Chandler, AZ
Inna Zakharevich	Henry M. Gunn High School	Palo Alto, CA

The USA team members were chosen according to their combined performance on the 31st annual USAMO and the Team Selection Test that took place at the Mathematics Olympiad Summer Program (MOSP) held at the University of Nebraska-Lincoln, June 18–July 13, 2002. Members of the USA team at the 2002 IMO (Glasgow, United Kingdom) were Daniel Kane, Anders Kaseorg, Ricky Liu, Tiankai Liu, Po-Ru Loh, and Tong-ke Xue. Titu Andreescu (Director of the American Mathematics Competitions) and Zuming Feng (Phillips Exeter Academy) served as team leader and deputy leader, respectively. The team was also accompanied by Reid Barton (Massachusetts Institute of Technology) and Steven Dunbar (University of Nebraska-Lincoln) as observers of the team leader, and Zvezdelina Stankova (Mills College) as observer of the deputy leader.

At the 2002 IMO, gold medals were awarded to students scoring between 29 and 42 points (there were three perfect papers on this very difficult exam), silver medals to students scoring between 23 and 28 points, and bronze medals to students scoring between 14 and 22 points. Loh's 36 tied for fourth place overall. The team's individual performances were as follows:

Kane	GOLD Medallist	T. Liu	GOLD Medallist
Kaseorg	SILVER Medallist	Loh	GOLD Medallist
R. Liu	GOLD Medallist	Xue	Honorable Mention

In terms of total score (out of a maximum of 252), the highest ranking of the 84 participating teams were as follows:

China	212	Taiwan	161
Russia	204	Romania	157
USA	171	India	156
Bulgaria	167	Germany	144
Vietnam	166	Iran	143
Korea	163	Canada	142

The 2002 USAMO was prepared by Titu Andreescu (Chair), Zuming Feng, Gregory Galperin, Alexander Soifer, Richard Stong and Zvezdelina Stankova. The Team Selection Test was prepared by Titu Andreescu and

Zuming Feng. The MOSP was held at the University of Nebraska-Lincoln. Because of a generous grant from the Akamai Foundation, the 2002 MOSP expanded from the usual 24–30 students to 176. An adequate number of instructors and assistants were appointed. Titu Andreescu (Director), Zuming Feng, Dorin Andrica, Bogdan Enescu, Chengde Feng, Gregory Galperin, Razvan Gelca, Alex Saltman, Zvezdelina Stankova, Walter Stromquist, Zoran Sunik, Ellen Veomett, and Stephen Wang served as instructors, assisted by Reid Barton, Gabriel Carroll, Luke Gustafson, Andrei Jorza, Ian Le, Po-Shen Loh, Mihai Manea, Shuang You, and Zhongtao Wu.

4 2001 Olympiad Results

The top twelve students on the 2001 USAMO were (in alphabetical order):

Reid W. Barton	Arlington, MA
Gabriel D. Carroll	Oakland, CA
Luke Gustafson	Breckenridge, MN
Stephen Guo	Cupertino, CA
Daniel Kane	Madison, WI
Ian Le	Princeton Junction, NJ
Ricky I. Liu	Newton, MA
Tiankai Liu	Saratoga, CA
Po-Ru Loh	Madison, WI
Dong (David) Shin	West Orange, NJ
Oaz Nir	Saratoga, CA
Gregory Price	Falls Church, VA

Reid Barton was the winner of the Samuel Greitzer-Murray Klamkin award, given to the top scorer on the USAMO. Reid Barton, Gabriel D. Carroll, Tiankai Liu placed first, second, and third, respectively, on the USAMO. They were awarded college scholarships of $15000, $10000, $5000, respectively, by the Akamai Foundation. The Clay Mathematics Institute (CMI) award, for a solution of outstanding elegance, and carrying a $1000 cash prize, was presented to Michael Hamburg for his solution to USAMO Problem 6.

The USA team members were chosen according to their combined performance on the 30th annual USAMO and the Team Selection Test that took place at the MOSP held at the Georgetown University, June 5–July 3, 2001. Members of the USA team at the 2001 IMO (Washington, DC, United States of America) were Reid Barton, Gabriel D. Carroll, Ian Le, Tiankai Liu, Oaz Nir, and David Shin. Titu Andreescu (Director of the

American Mathematics Competitions) and Zuming Feng (Phillips Exeter Academy) served as team leader and deputy leader, respectively. The team was also accompanied by Zvezdelina Stankova (Mills College), as observer of the deputy leader.

At the 2001 IMO, gold medals were awarded to students scoring between 30 and 42 points (there were 4 perfect papers on this very difficult exam), silver medals to students scoring between 19 and 29 points, and bronze medals to students scoring between 11 and 18 points. Barton and Carroll both scored perfect papers. The team's individual performances were as follows:

Barton	Homeschooled	GOLD Medallist
Carroll	Oakland Technical HS	GOLD Medallist
Le	West Windsor-Plainsboro HS	GOLD Medallist
Liu	Phillips Exeter Academy	GOLD Medallist
Nir	Monta Vista HS	SILVER Medallist
Shin	West Orange HS	SILVER Medallist

In terms of total score (out of a maximum of 252), the highest ranking of the 83 participating teams were as follows:

China	225	India	148
USA	196	Ukraine	143
Russia	196	Taiwan	141
Bulgaria	185	Vietnam	139
Korea	185	Turkey	136
Kazakhstan	168	Belarus	135

The 2001 USAMO was prepared by Titu Andreescu (Chair), Zuming Feng, Gregory Galperin, Alexander Soifer, Richard Stong and Zvezdelina Stankova. The Team Selection Test was prepared by Titu Andreescu and Zuming Feng. The MOSP was held at Georgetown University, Washington, DC. Titu Andreescu (Director), Zuming Feng, Alex Saltman, and Zvezdelina Stankova served as instructors, assisted by George Lee, Melanie Wood, and Daniel Stronger.

5 2000 Olympiad Results

The top twelve students on the 2000 USAMO were (in alphabetical order):

David G. Arthur	Toronto, ON
Reid W. Barton	Arlington, MA
Gabriel D. Carroll	Oakland, CA

Kamaldeep S. Gandhi	New York, NY
Ian Le	Princeton Junction, NJ
George Lee, Jr.	San Mateo, CA
Ricky I. Liu	Newton, MA
Po-Ru Loh	Madison, WI
Po-Shen Loh	Madison, WI
Oaz Nir	Saratoga, CA
Paul A. Valiant	Belmont, MA
Yian Zhang	Madison, WI

Reid Barton and Ricky Liu were the winners of the Samuel Greitzer-Murray Klamkin award, given to the top scorer(s) on the USAMO. The Clay Mathematics Institute (CMI) award was presented to Ricky Liu for his solution to USAMO Problem 3.

The USA team members were chosen according to their combined performance on the 29th annual USAMO and the Team Selection Test that took place at the MOSP held at the University of Nebraska-Lincoln, June 6–July 4, 2000. Members of the USA team at the 2000 IMO (Taejon, Republic of Korea) were Reid Barton, George Lee, Ricky Liu, Po-Ru Loh, Oaz Nir, and Paul Valiant. Titu Andreescu (Director of the American Mathematics Competitions) and Zuming Feng (Phillips Exeter Academy) served as team leader and deputy leader, respectively. The team was also accompanied by Dick Gibbs (Chair, Committee on the American Mathematics Competitions, Fort Lewis College), as observer of the team leader.

At the 2000 IMO, gold medals were awarded to students scoring between 30 and 42 points (there were four perfect papers on this very difficult exam), silver medals to students scoring between 20 and 29 points, and bronze medals to students scoring between 11 and 19 points. Barton's 39 tied for 5th. The team's individual performances were as follows:

Barton	Homeschooled	GOLD Medallist
Lee	Aragon HS	GOLD Medallist
Liu	Newton South HS	SILVER Medallist
P.-R. Loh	James Madison Memorial HS	SILVER Medallist
Nir	Monta Vista HS	GOLD Medallist
Valiant	Milton Academy	SILVER Medallist

In terms of total score (out of a maximum of 252), the highest ranking of the 82 participating teams were as follows:

China	218	Belarus	165
Russia	215	Taiwan	164
USA	184	Hungary	156
Korea	172	Iran	155
Bulgaria	169	Israel	139
Vietnam	169	Romania	139

The 2000 USAMO was prepared by Titu Andreescu (Chair), Zuming Feng, Kiran Kedlaya, Alexander Soifer, Richard Stong and Zvezdelina Stankova. The Team Selection Test was prepared by Titu Andreescu and Kiran Kedlaya. The MOSP was held at the University of Nebraska-Lincoln. Titu Andreescu (Director), Zuming Feng, Razvan Gelca, Kiran Kedlaya, Alex Saltman, and Zvezdelina Stankova served as instructors, assisted by Melanie Wood and Daniel Stronger.

6 2000–2004 Cumulative IMO Results

In terms of total scores (out of a maximum of 1260 points for the last five years), the highest ranking of the participating IMO teams is as follows:

China	1086	India	702
Russia	987	Belarus	700
USA	951	Iran	698
Bulgaria	942	Ukraine	694
Korea	843	Kazakhstan	643
Vietnam	842	Turkey	633
Taiwan	770	Israel	632
Romania	744	Germany	625
Hungary	716	Canada	605
Japan	705	Australia	553

More and more countries now value the crucial role of meaningful problem solving in mathematics education. The competition is getting tougher and tougher. A top ten finish is no longer a given for the traditional powerhouses.